ELECTROCHEMISTRY

ELECTROCHEMISTRY

A CONSISTENT THEORETICAL TREATMENT

Harvey N. Seiger

Writers Club Press

San Jose New York Lincoln Shanghai

Electrochemistry
A Consistent Theoretical Treatment

Writers Club Press
an imprint of iUniverse, Inc.

For information address:
iUniverse, Inc.
5220 S. 16th St., Suite 200
Lincoln, NE 68512
www.iuniverse.com

ISBN: 0-595-19495-8

Printed in the United States of America

Contents

Introduction

"Anion electrodes are best obtained by means of a metal in contact with one of its insoluble salts immersed in a solution of a soluble salt of the same anion, e.g., $Hg|Hg_2Cl_2(s)$ in KCl solution. If current is passed from electrode to electrolyte, mercury dissolves to form mercurous ions; these unite with chloride ions in the solution to form insoluble mercurous chloride, and the result is the removal of chloride ions from solution. On passing current in the opposite direction mercurous ions are discharged, the solution becomes unsaturated and mercurous chloride dissolves; the result is that chloride ions have passed into solution. The system thus behaves as a reversible chlorine electrode, for the chemical changes can be made to occur in either direction by infinitely small changes of applied external E.M.F."

The above quotation is copied directly from Glasstone's "Physical Chemistry"[1] page 923. Undergraduate chemistry teaches that oxidation occurs at the anode and, conversely, reduction takes place at the cathode. We need such a definition because secondary batteries, as is the common automotive lead acid battery, do not change polarity between normal "charge" and "discharge" processes. The active material, lead, is present in the material of the negative electrode and becomes an insoluble lead sulfate during "discharge". Simultaneously, the lead dioxide material in the positive electrode "discharges" also to lead sulfate.

An external power supply is required to "recharge" this battery. If one has a voltmeter attached to the terminals of a lead acid battery one reads about 2.1 volts per cell on open circuit, something less than 2.1 volts during "discharge", and something more than 2.1 volts during the subsequent "recharge"

The quotation from Glasstone contains a fallacy and let us see what this fallacy is and what it means. In his first step he lets mercury change to mercurous ion. To have such a change requires removal of an electron into an external circuit, and also requires second electrode that picks up this electron and causes some reductive process. We are not through yet, to complete this electrochemical circuit, the charge is carried between the two electrodes in the solution by ions. At each electrode there is a charge transfer process accounting for the change in the charge carrier from ions to electrons (or holes) at one, and the reverse at the other. If we now look at the process in the quote we find that the mercury/mercurous change is an oxidation and this is the anode. The chloride ion does not undergo a valence change so this is not a chlorine electrode contrary to the statement in the quotation that "this behaves as a reversible chlorine electrode". The electrode behaves as a $Hg/Hg^{+,}$ KCl electrode.

Is it important to investigate such fallacies? In 1983 this writer reviewed a proposal for a lithium/fluorine battery expected to display about 6 volts. The author of the proposal in effect took the $Hg/Hg_2Cl_2,KCl$ electrode and substituted Ag/AgF, molten salt containing fluoride. So, while 6.1 volts per cell sounds appealing, 3.9 volts does not have the same appeal. It is interesting to note the time span involved between book publication and the date of the proposal.

If we now ask the question about the existence of problems in electrochemistry, we find a number of them. Guggenheim[2] promoted the concept that single electrode potentials can not be used, and further, there was an absence of absolute single electrode potentials. I concur with Guggenheim, and go even further to show that if one employs single electrode potential concepts some ideas are developed that are not physical and are misleading instead of being useful. The foundation for the treatment of absolute potentials has already been given with the concept of an electrochemical circuit.

Batteries, particularly secondary batteries, are fast electrochemical reactions. Yet, no treatment fast electrochemical reactions are found.

This subject is also treated in this book. Other electrochemical processes are investigated critically, including some that deal with membrane phenomena and reach into electrochemistry and medicine.

1. S. Glasstone, *Textbook of Physical Chemistry*, Second Edition, D. Van Nostrand, 1946.
2. E.A. Guggenheim, J. Phys.Chem.,3-44,1758 (1930), Trans. Faraday Soc. *36*, 139, (1940).

I

Absolute Potentials

Electrochemical processes are a subclass of chemical reactions in which an oxidation process occurs at the anode and a reduction process which is coupled to this anode and, therefore, simultaneously occurs at the cathode. For these coupled processes to take place there is an electronic path outside the cell and an ionic path within the cell which, of course, lies between the two electrodes. When the reaction is spontaneous the Gibbs energy for the oxidation-reduction reaction has a negative value. Such spontaneous processes are (1) batteries and fuel cells which deliver energy to a load and, (2) many systems undergoing corrosion. The Gibbs energy is the maximum work that can be done and, in practice, as we already know, is never achieved.

If the electrochemical reaction is not spontaneous and energy must be supplied from an external source the processes are still galvanic, in the case of inert electrodes in an ionically conductive solution, electrolysis takes place. In the case of an electrolyte containing appropriate cations electroplating takes place. In a third case, that of secondary batteries, the electrode chemicals are regenerated resulting in storage of chemical energy, until the stoichiometrically limited quantities are consumed and the battery enters "overcharge" with a concurrent change in electrode process usually at one of the two electrodes.

If we use what we have we have written above, the anode process may be written as:

(1) $M \rightarrow M^{+2} + 2e^-$.

The cathode process captures these electrons and the active material is reduced. Let us exemplify this with an oxide and use an aqueous alkaline solution as the electrolytic solution, then we may write:

(2) $BO + H_2O + 2e^- \rightarrow B + 2\,OH^-$ ·

Adding more detail and further assuming that the M^{++} ion hydroxide or oxide is insoluble, the first equation is modified to become:

(1') $M + 2OH \rightarrow M(OH)^2 + 2e^-$.

If equations 1' and 2 are summed the OH^- ions cancel along with charge, but for our purposes these equations are better left in the present form. In this form we see that the OH^- ion flow in the electrolyte carries charge and that electrons, or the equivalent, flow through the external circuit, and that the charge transfer is from ionic species to electronic conductor. The charge transfer is contemporaneous with the change of oxidation state; further, the current flows as a continuous circuit through the cell so that while oxidation takes place at the anode, reduction simultaneously occurs at the cathode.

When the case where the above reactions are spontaneous, they represent a battery (or fuel cell) and the potential is given by the relationship for Gibbs energy:

(3) $-\Delta G = zFE$

If the system delivers energy into the external circuit, by convention, the Gibbs energy is negative. Under these circumstances the battery cell voltage has a positive value. When the battery is capable of delivering energy into the load the anode is the negative terminal and the cathode the positive terminal. In galvanic systems that are not spontaneous, such as electrolysis cells and plating processes, the signs are opposite to batteries. Continuing with spontaneous processes, when the stoichiometric quantities in an electrode are consumed, the electrode process changes and the polarity of the terminals may change. The battery cells are of further interest because overcharge results in electrolysis without a change of polarity while the measured voltage changes to some degree associated with the process change from the normal charge process to the overcharge process. Overdischarge, on the other hand, usually results in a change in magnitude of the voltage as well as polarity. There are instances where overdischarge does not cause polarity to reverse such as batteries with lithium anodes in which the stoichiometric quantity of lithium exceeds that of the active cathode material. The lithium has such a negative potential that the second process at the cathode is still positive with respect to the lithium electrode. Similar situations can occur with corrosion processes.

The coupling of the anode and cathode processes means that the desired reaction can take place only with a complete circuit. Break the external path between the anode and cathode and the electrons cannot leave the anode. There is no circuit. Similarly, break the internal path between the electrodes so that ions can not flow between the electrodes, and the electrochemical processes no longer take place. Again, there is no circuit. There is, instead, an oxidizing agent without chemicals to be oxidized that is isolated from a reducing agent that also can not react. These facts lead to the conclusion that *two* electrodes are required for electrochemical process to occur. Hence, there really are no single electrodes. When electrochemists want to measure "single electrode potentials" they use a reference electrode and a system containing (1) an oxidized material, (2) a conjugate reduced material, (3) an electronic conductor and an ionic conductor in the proper locations. While the composition of several different kinds of reference electrode may be differ, they still contain the four subsystems, but the ionic conductors are connected internally and the electronic conductors are connected externally. A measurement of voltage difference is then made using a device that, ostensibly, does not draw current. In the past potentiometers had been used, but this practice has been replaced by high impedance electronic meters. Any measurement made is not an open circuit measurement, but one in which the amount of current passing through the system is assumed to be negligibly small. However, as we have just discussed unless there is an electronic connection there is no circuit.

Even in this case there are no single electrodes, only pairs even when one electrode is common to two others.

Since there are no single electrodes but only pairs of galvanic couples, the voltages measured are differences in potential of the two sets. To make this measurement meaningful, one system is selected and defined as the base reference electrode. That is, the hydrogen electrode defined as hydrogen gas at a fugacity of unity in an acidic electrolyte with protons at unit activity at 25°C; this arrangement is arbitrarily assigned a value of zero. Now all reference electrode measurements are based upon this and the thermodynamic principle that the energy content is dependent upon the state of the system and not the path by which it arrived at that state. This is far-reaching since almost any sort of reference electrode can be used providing the process taking place can be inferred even if not empirically measured.

Absolute electrode potentials should really be related to the study of the kinetics of a complete system having oxidizers and reducing agents and the external and internal connections. In this instance the results will be the potential difference between the electrodes which is related to the Gibbs energy. Because the processes that take place at electrodes may be different than if there is a direct reaction, the Gibbs energy values may differ from the values obtained by other methods, say thermochemically. An example of this is the hydrogen-oxygen reaction in which the thermochemical inferred value for ΔG corresponds to 1.229 volts and fuel cells measure just about 1.0 volt even when the current passing through the cell is so small as to be considered negligible.

Galvanic cells, particularly battery cells, are frequently represented in circuits as parallel plates of unequal size. This simple representation overlooks the four subsystems for an electrode and any representation of a charge transfer interface. An improvement may be made by allowing some volume to indicate one oxidation state, a second volume for the other state, another volume for the electrolyte and finally a line to indicate the electronic conductor. The electrolyte volume is shared by the two electrodes and the galvanic system is diagrammed as shown in Figure I-1. This becomes a galvanic cell when the two electronic conductors are connected to complete the circuit

Figure I-1. Diagram of a galvanic cell showing the two oxidized regions, the two reduced regions and the separator.

The rectangle on the left represents the oxide and the one between it and the electrolyte, the clear rectangle, is the reduced state of the cathode active material. Similarly, the rectangle on the right represents the reduced state of the anode; the rectangle situated between it and the electrolyte is the oxidized state of the anode active material.

There is an order to this arrangement - highly oxidized material, the reduced form, electrolyte, oxidized material, and finally the reduced active material. The electrochemical process of the anode occurs at an interface between the active material and the electrolyte with the oxidized metal depositing near the interface. The reduction reaction of the cathode also occurs at an interface between the highly oxidized material and the electrolyte. The electronic conductors are not directly

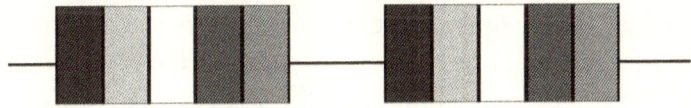

Figure I-2. Representation of a two cell battery.

connected to the interfaces and are not at the same potential. This can be proved by considering a two cell battery as shown in Figure I-2. The purpose of putting two cells in series is to make a battery with twice the voltage of a single cell, a globally used concept where n cells in series yields an open circuit voltage that is the sum of all n included cells.

Understanding that when a single electrode potential is mentioned below it means with respect to a reference electrode, let the voltage of the cathode in cell 1 be V_{c1} and in cell 2 is V_{c2} while the anode voltages are respectively V_{a1} and V_{a2}. We need now to prove that V_{c2} [1] V_{a1} when they share the intercell connector. The nature of the proof is reductio ad absurdum where it shall be assumed that the electrical connection between them causes the cathode of cell 2 to be at the same potential as the anode of cell 1 which may be written as $V_{c2} = V_{a1}$; this has been termed by some as the "equipotential" hypothesis. The battery voltage is equal to the sum of the voltages of cells 1 and 2, $V_{bat} = V_{cell1} + V_{cell2}$. The cell voltages are the difference in potential between the cathode and the anode in each one, or $V_{cell1} = (V_{c1} - V_{a1})$ and $V_{cell2} = (V_{c2} - V_{a2})$. Now, substitute the single electrode values into the equation for V_{bat} obtaining $V_{bat} = (V_{c1} - V_{a1}) + (V_{c2} - V_{a2})$; which may be rewritten as $V_{bat} = V_{c1} + V_{c2} - V_{a1} - V_{a2}$. Invoking the so-called equipotential assumption, $V_{c1} = V_{a1}$ the last equation becomes $V_{bat} = V_{c1} - V_{a2}$ which is the wrong answer. The equipotential hypothesis is rejected and it is recognized that the source of the potentials measured is at the respective electrode/electrolyte interfaces. Furthermore, each electrode/electrolyte interface has the active species in the two different oxidation states.

When the above argument was pointed out to an advocate of the "equipotential" hypothesis he countered by pointing out that a voltmeter placed across the terminals of a n cell battery shows the voltage of all n cells. Too bad he didn't have a second voltmeter to probe all the cells of the battery so he could prove for himself that each cell still had its characteristic voltage. Since one can't have it both ways, the "equipotential" hypothesis has to be abandoned.

II

Electrochemical Kinetics

The first chapter applied thermodynamics to the electrodes with some consideration of the mechanism. Even when loads or measuring instruments draw infinitesimally low currents the two electrodes are connected through the external circuit thereby completing the overall electrochemical circuit. The thermochemical and electrochemical values for the Gibbs energy may differ because the reactants are ostensibly the same but the products may be different. They may differ in composition such as water of hydration or crystalline form. Whether we consider batteries, or electrolysis, or electroplating, significant currents are used and chemical, or better, electrochemical kinetics become important. Chemical and electrochemical kinetics differ merely by consideration of the charge passing through the electrode.

Early in this century Tafel[1] found a logarithmic relationship between the current passing through some galvanic cells and the voltage differential applied to the cells. The voltage differential is termed the overvoltage. Sometime after the quantum theory was established Gurney[2] derived an equation that yielded the same kind of logarithmic relationship between current through some galvanic cells and the overvoltage. The theory was applied to single electrodes without consideration given to the counter electrodes. This was the situation established in the first chapter to have no meaning since it was not a galvanic cell. When this Gurney equation was applied there evolved two energy barriers to the system, one at each interface. If we redo the derivation, but recognize only one potential energy barrier between the reactants and products an expression is derived which is similar in form but has a different physical interpretation. The Tafel[1] type relationship is related to activation control. There are four kinds of effects that can control electrochemical reaction rates. The first can be the activation energy barrier. The second possible rate controlling mode can be diffusion of a reactant to the interface or the diffusion away from the interface of a reaction product. The third possible rate controlling step is ohmic in nature. These three effects are distinguishable since activation control demonstrates what appears to be a logarithmic relationship between current and voltage, concentration overpotential decreases with stirring, and ohmic effects show a linear relationship between voltage and current. The fourth type is a special class of ohmic potential dealing with reactions that can occur very rapidly and chemical kinetics does not treat them because the activation energy barriers are so low that the reaction may even be explosive.

Some electrochemical reactions are fast reactions. When lead acid batteries are used for welding, and nickel cadmium and some high temperature batteries are used for pulse power the discharge rates are very fast. Lithium/silver oxide batteries explode when there are short circuits. The so-called super- and ultracapacitors are also electrochemical systems that are capable of having energy removed in very short time intervals. These reactions are so fast that they can not be handled by ordinary chemical kinetics which deal with molecular collisions and are relatively slow reactions. When we deal with rapid reactions the rate limiting process may be the magnitude of the external load. The battery voltage decreases due to losses that occur within the electrochemical system. Such losses may be due in part to the conductivity of the electrolyte and governed by the pathway along which conduction occurs within the cell. The electrolyte conductivity will manifest itself as an ohmic resistance as long as we deal with direct current and not with alternating currents, particularly at high frequencies.

The electrochemical reaction starts at an interface where the impedance to the reaction is minimal. The reaction front or boundary moves with time as the reactants become exhausted locally. The impedance to the reaction increases with time and is affected by current density. To gain some insight into the relationship between current and voltage for these fast reactions we shall model the phenomenon using an array of concentric cylinders.

a. Moving Reaction Zone Treatment.

For modeling purposes the active material and the electrolyte were treated as an array of concentric cylinders with the active material surrounding the electrolyte. We had previously investigated a high rate discharge of the AgO electrode under flowing electrolyte conditions. The silver electrode has unique properties for such an investigation because AgO is gray, Ag_2O is black, and Ag is white. Interrupting the discharge at various states of charge showed that the discharge at high rates went directly from AgO to Ag without evidence of Ag_2O, and that the reaction proceeded from the front surface back to the collector, a direction which we shall referred to as the z axis. The reaction front had a slight wave shape, nearly straight. The very front of the electrode was still partly dark, mottled in appearance, and the mottling did not disappear until the discharge was completed. By analogy, it can be reasoned that the front to back discharge is applicable to all electrochemical systems. Further, there were no dimensional changes at all in the x and y directions of the AgO electrode, and very little dimensional change in the thickness. While the discharge process of the AgO leads to an increasing porosity because of relative molar volume changes, other systems might undergo a decrease in porosity as a result of the electrochemical reaction. Such a decrease in porosity is expected to result in stronger reasons for front to back discharge.

Thus it can be assured that the reaction occurs in a zone favorable to it at the front and then moves inward giving us a moving reaction zone.

The conduction in the thin layer of active material and in the electronic pathway to the collector may decrease as the reaction zone moves back, but the ionic pathway in the pore increases and this can be shown to be a more major contributor to voltage drop with decease of state of charge. The purpose of this work is to use this volume change to evaluate the change of voltage with time and current during discharge.

The active material is concentric to the electrolyte having a radius R assumed to remain constant during the discharge, based on the findings for the Ag electrode. The ionic pathway resistance is taken as a reaction zone length λ' multiplied by the dimensions of the ionic path. The front surface is unique because some of the electrolyte is allocable to the first zone for discharge. This additional electrolyte is seen as λ^* in Figure II-1. We shall assume that the reaction zone is $\lambda'-\lambda$ and the available capacity in each zone is: $\left\{ \pi \left(R^2 - r_o^2 \right) \left(\lambda' - \lambda \right) d_c \right\} \varphi$ where d_c is the density of the active material in g/cm³ and φ is the conversion factor from mass to ampere-hours. The electrolyte within the pore starts with a radius r_o and penetrates the pore length λ.

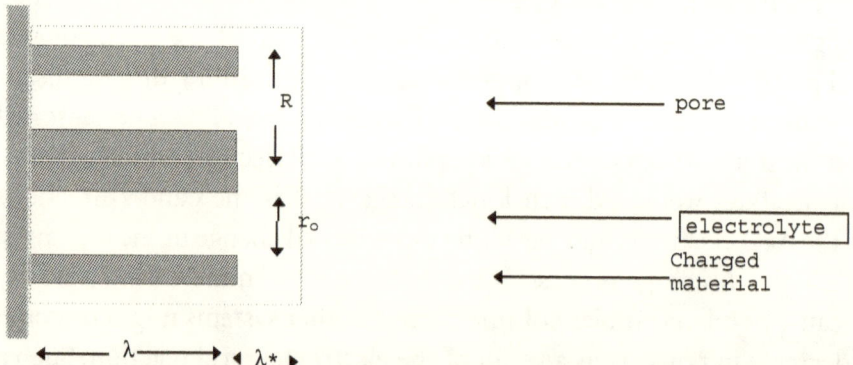

Figure II-1. Concentric Cylinder Model of a Porous Electrode

The volume change per ampere-hour of discharge is determined as follows. The molecular weight of the charged active material divided by the density of this material yields the molar volume. Similarly, for the discharged state, division of the molar volume divided by the density of the discharged material yields its molar volume. If we divide the change in the molar volume of one mole of electroactive material then we

obtain the change in volume per coulomb:
$$\left[\left(\frac{W_d}{d_d} - \frac{W_c}{d_c}\right) / zF\right] = \frac{K}{zF},$$

$$K = \frac{mW_c}{d_c} - \frac{mW_d}{d_d}.$$

where we conveniently let

The number of reaction zones encountered is $\lambda/(\lambda'-\lambda)$. The ionic impedance presented by each exhausted zone is $\rho_e(\lambda'-\lambda)/\pi r^2$. When this is multiplied by the pore current, j/n, the voltage drop is obtained. The contribution to the voltage drop due to pore impedance, ΔV_{pore} increases from zero at the start of discharge to:

$$\frac{\lambda}{(\lambda'-\lambda)} \times \rho_e \times \frac{(\lambda'-\lambda)}{\pi r^2} \times \left(\frac{j}{n}\right) = \Delta V_{pore} = \rho_e\left(\frac{\lambda}{\pi r^2}\right) \times \frac{j}{n},$$ at the completion of the discharge and the increase is linear with the state of charge.

Digressing, we want to establish the relationship between the way r changes with current and time. The initial volume of electrolyte, v_i, in a pore is: $v_i = \pi r r_o^2 \lambda$. The change in volume of the pore is:

molar volume charged - molar volume discharged,
which, considered on the basis of the discharge is:

$$\left(\frac{W_c}{d_c} - \frac{W_d}{d_d}\right) / zF = K\frac{cm^3}{coulomb},$$ where z is the valence change and F is the faraday constant (96500 coulombs per equivalent).

The capacity associated with each pore is given by:

$$\left(\pi R^2 - \pi r_o^2\right)\lambda\frac{d_c}{w_c}zF = \qquad Capacity \text{ (in coulombs)}$$

The pore volume at the end of discharge is:

$$v_f = \pi r^2 \lambda \quad \text{and}$$

$$v_f - v_i = \pi r^2 \lambda - \pi r_o^2 \lambda.$$

Recognize that jt = the capacity discharged in coulombs. If there are n pores/cm^2, then the capacity associated with a pore is:

$$\frac{jt}{n} \text{ coulombs cm}^3/\text{coulombs. Now:}$$

initial volume - $\dfrac{jtK}{n}$ = final volume. Substituting:

$$\pi r_o^2 \lambda - \frac{jtK}{n} = \pi r^2 \lambda \qquad \text{, or:}$$

$$r = r_o^2 - \frac{jtK}{\pi\lambda n} \text{ , whence}$$

$$r = \left(r_o^2 - \frac{jtK}{\pi\lambda n}\right)^{\frac{1}{2}}$$

If a reaction zone has a thickness $\left(\lambda'-\lambda\right)$, the number of reaction zones encountered is $\dfrac{\lambda - \lambda'}{\pi r^2}$. The ionic impedance by each zone is $\rho_e\dfrac{\lambda - \lambda'}{\pi r^2}$. Multiplying this by the pore current, the voltage drop is obtained. The contribution to the voltage drop due to pore impedance increases from zero initially to:

$$\frac{\lambda}{\lambda'}\rho_e\frac{\lambda'-\lambda}{\pi r^2}\left(\frac{j}{n}\right)=\Delta V_{pore}$$

$$\Delta V_{pore}=\rho_e\frac{\lambda}{\pi r^2}\left(\frac{j}{n}\right)$$, at the completion of the discharge and is

linear with state of charge. Using the separately evaluated functional relationship between r and r° the following is now obtained:

$$\Delta V_{pore}=\frac{\lambda}{\pi}\frac{j}{n}\times\frac{1}{r_o^2-\frac{jtK}{\pi\lambda n}}$$ which simplifies to:

$$\Delta V_{pore}=\frac{\rho_e\lambda^2 j}{\pi\lambda n r_o^2}$$

For short periods of discharge and at low current density the voltage drop increases linearly with time. For high current densities and for longer discharge periods the curvature appears in the voltage drop equation. This kind of behavior has been found in nickel-iron batteries[3] and in pulse type of discharges from nickel-cadmium batteries[2].

1   J. Tafel, Z. Physik. Chem., *50*, 641 (1905).
2   R. W. Gurney, Proc. Roy. Soc.,A134, 137 (1932).
3   U. Falk and A.J. Salkind, *Alkaline Storage Batteries*, J. Wiley& Sons, Inc. New York, 1969

III

Rate Considerations in Electrochemical Reactions

The fast reactions had no potential energy barrier and the rates would be instantaneous except for the impedance external to the circuit as well as within the electrochemical device. Now we want to consider slower fast reactions had no potential energy barrier and the processes of the nature usually the subject of chemical kinetics. The slow processes may have rate limiting steps due to transport process or due to having a potential energy barrier so that individual species must statistically have sufficient energy to surmount the barrier or to tunnel through it. The limitations due to transport processes are considered diffusion controlled reactions while those having a potential energy barrier are called activation controlled reactions[1]. At this point we shall treat the activation controlled reactions.

a. Rate Limitation Due to An Activation Energy Barrier.

The treatment of fast electrochemical reactions given previously is new since no similar approaches had been found in the literature. There are many treatments using the potential energy barrier in the literature. The Tafel equation[2] is characteristic, but Tafel's 1907 work was experimental and it remained for Gurney[3] to develop a derivation using quantum mechanical considerations. Subsequent to this original work were modifications by Butler[4] and by Volmer[5.] All of these kinetic equations yielded linear equations when they were expanded about zero

14

overvoltage using the Maclauren expansion and an exponential form for values of overpotential significantly different from zero. Since these two results are consistent with experimental data for some electrolysis and some plating processes the work has been accepted.

Subsequent to Gurney another investigator, Guggenheim[6] reasoned that there are no single electrodes. Earlier we had shown our agreement with Guggenheim and have always used a pair of canonical electrodes. Further, we have established that a single electrode potential is meaningless, but that measurement of electrodes with respect to a defined reference electrode is meaningful provided we can reasonably establish that the electrode process at the reference electrode is invariant. At very low currents we may invoke the Onsager Principle of Microscopic Reversibility[7] providing we do so cautiously, meaning an absence of a potential step or other singularity.

When we treat of an electrochemical reaction, a battery or electrolysis cell or a corrosion cell or an electrodeposition device, the currents are of significant values. Further, the current has a solenoidal path so that if the electrode process at one electrode is oxidative, the electrode process at the other is reductive. The Gibbs energy is given by the well-known equation, $-\Delta G = z\Phi E$, and the chemical reaction involved is represented by:

$$BO + M \rightarrow MO + B + Q$$

where Q is the energy change accompanying by the reaction. If the reaction is spontaneous no external energy source is required which is the situation with the battery and corrosion cell cases. If the reactions are forced to occur as in electrolysis and electrodeposition then they require an external electrical source of energy. Let us treat this latter situation wherein energy must be supplied to make the reaction proceed (reenergizing a secondary battery is also included). In this case there is a potential energy barrier and the applied potential is assumed to alter the ground state energy level. The reactants in these examples are similar to those of the generalized reaction written above. Using two different situations we can further exemplify the electrode processes:

cell   Cu|CuSO$_4$ solution | Cu, substrate

anode: Cu Cu^{++} + 2e$^-$

cathode: Cu^{++} + 2e$^-$ → Cu (substrate) ,

which is an electrochemical plating process, and

cell: H$_2$O →1/2 O$_2$ + H$_2$,

   anode: 2OH$^-$ → H$_2$O + 1/2O$_2$ + 2e$^-$

cathode: 2H$_2$O + 2e$^-$ → H$_2$ + 2OH$^-$

which is an example of an electrolysis process. The latter reaction is particularly interesting since hydrogen and oxygen are the components of one kind of fuel cell or battery. If the energy supplied to these cells raises the energy level of the ground state of the reactants then the effective

energy barrier is $F^a - z\Phi\delta V$, where the first term is the activation energy barrier and the variation represents the applied voltage difference and is also the decrease in height of the energy barrier.

From the several examples above it is noted that the net electrochemical reactions are the same as the chemical reactions as well they should be. The charge is transported electronically in the external circuit and ionically in the internal circuit. There are two important interfaces. The anode sends electrons into the external circuit and simultaneously interacts with the ionic charge carrier in the electrolyte. This charge transfer interface results in oxidation. Similarly, at the cathode there is a charge transfer interface wherein the electrons are received and interaction at the interface results in reduction of the active material and production of the ionic charge carrier of the electrolyte. Because these reactions are coupled there is but a single potential energy barrier and not two or more. This point is stressed because of the departure from, collectively, Gurney , Butler and from Volmer because of this change in concept. It is important to note that with this kind of treatment certain concepts in the prior literature are consequently changed. For instance, the concept of an exchange current density does not apply nor does the concept of the "Transfer Coefficient" which says a fraction of the overpotential is used to raise one side of the potential energy barrier and the remaining fraction is used to lower the other side of the barrier. Instead, the external circuit behaves as if the ground state of the reactants undergo a change in level and the system is now at an energy level represent by V^T. In a later chapter we shall compare the concept of an exchange current density with certain isotopic exchange reactions.

The potential energy barrier is shown in Figure III-1. We assume that the voltage applied to the system affects just one side of the barrier; it raises the energy level of the reactants without affecting the products. While Figure III-1 represents the reactants as being in the lowest energy state, the applied voltage can decrease the height of the energy barrier so that the reaction can proceed and reach an energy state higher than the ground state. The Gibbs energy is stored in the products. In this way electroplating and electrolysis can take place. This sort of reasoning also helps explain the recharge of batteries and fuel cells. The required energy is greater because of reversible and irreversible entropic changes. Whenever the term "recharge" is used it is considered a colloquialism for energy storage and not for coulombic storage. The energy is stored as chemical energy and the electrochemical device is just a means for conversion of chemical to electrical energy.

Using the Law of Guldberg and Waage we may write the kinetic equations for the reaction rates:

$$-\frac{d(rcac\tan ts)}{dt} = k_f \pi C_i$$

$$\frac{d(products)}{dt} = k_r \pi Cj$$

which at equilibrium becomes:

$$\frac{k_f}{k_r} = \frac{\Pi C_i}{\Pi C_j}$$

Using Figure III-1 we may then write:

$$k_f = k_f^o \exp\left[-(F^a + \Delta G - z\phi\delta V)/RT\right]$$

and

$$k_f = k_f^o \varepsilon^{F^a/RT}$$

.

The current is the difference in rates of the forward and reverse reactions when the rates are appropriately converted using zF as the conversion factor, hence:

$$i = z\phi\left[k_f^o \varepsilon^{(F^a + \Delta G - z\phi\delta V/RT)} - k_r^o \varepsilon^{-F/RT}\right]$$

When no current is drawn $\delta V = 0$ and $i = 0$, hence:

$$0 = z\phi\left[k_f^o \exp-(F^a + \Delta G)/RT\right] - k_r^o \exp-(F^a/RT)$$

$$k_f^o \exp\left[-(F^a - \Delta G/RT)\right] = k_r^o \exp\left(-F^a/RT\right)_{k^o{}_f e^-}$$

$$\ln k_f^o - \frac{F^a}{RT} - \frac{\Delta G}{RT} = \ln k_r^o - \frac{F^a}{RT}$$

$$\frac{\Delta G}{RT} = \ln \frac{k_f o}{k_r^o} = \ln K$$

or

$$K = \exp(-\Delta G/RT)$$

and $-\Delta G = RT \ln K$.

This is the well-known thermodynamic relationship between Gibbs Energy and the equilibrium constant. The Gibbs Energy is related to the electromotive force by another well-known equation, $-\Delta G = z\Phi E$, and changes the concept of what happens with an applied voltage, either increasing or decreasing the energy level but does not change fundamental experimental results. The thermodynamic functions may be substituted into the net current equation

$$i = z\Phi[k_f^o\{(\exp-\left(F^a + \Delta G - z\Phi\delta V\right)/RT\} - k_r^o\exp\left(-F^a/RT\right)]$$

$$i = z\Phi(\exp- F^a/RT)\left(\frac{k_f^o}{K}\exp-\Delta G/RT)(\exp^{z\Phi\frac{\delta V}{RT}}) - k_r^o\right)$$

$$i = z\Phi\exp\left(-\frac{F^a}{RT}\right)k_r^o\left(\exp(\frac{z\Phi\delta V}{RT}) - 1\right)$$

The relationship between current and voltage is linear for small values of overvoltage. One set of experiments indicate the region of linearity is between zero and 15 mV for hydrogen deposition on a clean Ni surface[8]. The generalized relationship between current and voltage is that of an exponential when the applied voltage is greater that the activation energy term. A comparison between the equation just derived and those based on the previous concept shows that at values of overvoltage above about 0.1 volt there are no real differences, and that at very small overvoltages they tend to the same limit. It is only at values of overvoltages grater than about 15 mV to 100 mV are there any differences, and even these differences are not significant when applied to a reaction having a 14 kcal/mol activation energy.

b. Diffusion Controlled Electrochemical Reactions.

There are some electrochemical reactions that are inherently fast but are limited by the rate at which a reactant reaches the zone at which charge transfer occurs. Imagine a three electrode system consisting of a working electrode, a counter electrode and a reference electrode. Each participates in an oxidation/reduction reaction resulting in two combinations of interest, namely:

- the working electrode versus the counter electrode, and
- the working electrode versus the reference electrode.

Consider, first, the WE/RE cell. On open circuit there is a potential in which the reference electrode has an oxidized form and a reduced form. Suppose the RE is a standard hydrogen electrode where the oxidized form is H^+ at an activity a_H (unity) and the reduced form is gaseous hydrogen with a fugacity fH (also unity) and the corresponding electrochemical process is:

$$H_2 \rightarrow 2\,H^+ + 2\,e^-$$

The working electrode taken as an example is the I^-/I^{-3} system which is written as:

$$I^{-3} + 2\,e^- \rightarrow 3\,I^-.$$

The net reaction is the sum of two half cells, or:

$$I^-_3 + H_2 \rightarrow 3I^- + 2H^+,$$

and in the Nernst equation form we write:

$$E = E_o + \frac{RT}{zF}\ln\frac{(H^+)^2(I^-)^3}{(I^-_3)(H_2)}$$

$$E = E_o + \frac{RT}{zF}\ln\frac{a_H^2(I^-)^3}{f(I^-_3)} = E_o + \frac{RT}{zF}\left[\ln\frac{a_H^2}{f} + \ln\frac{(I^-)^3}{I^-_3}\right]$$

Since the current drain is small, a well designed reference electrode would have a substantially constant value of which can be lumped into E_o to give a new E', i.e.:

$$E = E' + \frac{RT}{zF}\ln\frac{(I^-)^3}{(I^-_3)}.$$

Let us arrange the experimental set up in such a way that there is an excess of I^- ion so that its concentration may be considered constant and we shall also make two assumptions as follows:

- the diffusion of I_3^- from the bulk solution through the convectodiffusive layer controls the activity (concentration) of I_3^- at the metal surface.
- the reduction of triiodide to iodide ions is so rapid that the electrode is always in equilibrium.

With these assumptions we write:

$$E = E' + \frac{RT}{zF} \ln \frac{const}{I_3^-} = E'' - \frac{RT}{zF} \ln\left(I_3^-\right)$$

The limiting current occurs when every triiodide that reaches the charge transfer interface is reduced to iodide ions. The flux of triiodide is given by:

$$\Psi = \frac{D}{\delta}\left[\left(I_3^-\right)_{bulk} - \left(I_3^-\right)\right]$$

and the limiting current is:

$$i_f = 2F\Psi = 2F\frac{D}{\delta}\left(I_3^-\right)$$

while the current in other cases is:

$$i = 2F\Psi = 2F\frac{D}{\delta}\left[\left(I_3^-\right)_{bulk} - \left(I_3^-\right)\right]$$

and dividing to obtain

$$\frac{i}{i_l} = 1 - \frac{\left(I_3^-\right)_{surface}}{\left(I_3^-\right)_{bu;k}},$$

rearranging:

$$\left(I_3^-\right)_{surface} = \left(I_3^-\right)_{bulk}\left(1 - \frac{i}{i_l}\right)$$

which may be substituted into the Nernst equation as:

$$E = E'' - \frac{RT}{zF} \ln \left[(I_3^-)_{bulk} \left(1 - \frac{i}{i_l} \right) \right] \text{, or}$$

$$E = E''' - \frac{RT}{zF} \ln \left(1 - \frac{i}{i_l} \right)$$

which is the relationship between current and voltage between the working electrode and a reference electrode. The reference electrode must be placed into the cell to avoid inclusion of voltage drops while measurements are made.

1   M. Shaw, Fundamentals of Electrochemistry, Report 3106, Whitaker, 1964.
2   J. Tafel, Z. Physik. Chem. 50, 641 (1905).
3   R.W. Gurney, Proc Roy. Soc. A134, 137 (1932).
4   J.A.V. Butler, Proc. Roy. Soc. A157, 423 (1936)
5   T. Erdy-Gruz and M. Volmer, Z. Physik. Chem. *150A*. 203, (1930)
6   E.A.Guggenheim,J.Phys.Chem.33,842,(1929); idem. ibid. 34, 1540, (1930).
7   S.R. De Groot, *Thermodynamics of Irreversible Processes, North Holland Publishing Co.*, Amsterdam

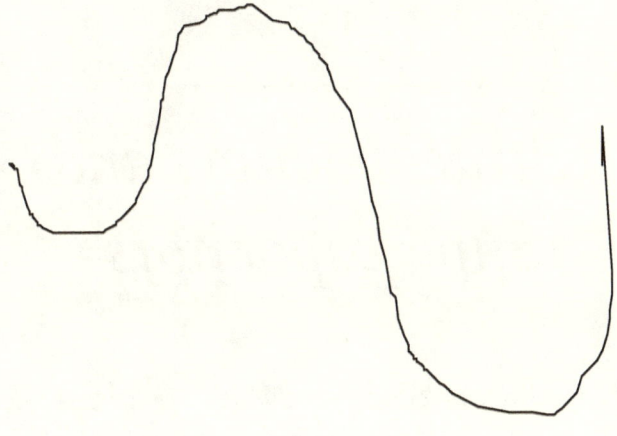

activated state

initial state

final state

Figure III-1. A potential energy barrier curve for a spontaneous reaction in which the energy level of the final state is lower than the energy level of the initial state.

IV

Ultracapacitors and Supercapacitors

One of the capacitors or condensers which we have studied in our undergraduate physics courses consisted of two metal plates separated by an insulator. The insulator is frequently termed the dielectric. The most notable feature of such condensers was that the voltage between the plates was proportional to the charge stored on those plates. The medium between the plates affects the amount of charge stored and the effect is used to define the dielectric constant. The relationship for parallel plate condensers is:

$$V = \frac{q4\pi}{kA}d$$

where q is the charge, k the dielectric constant, A the plate area and d the distance between the plates. It is assumed that the distance between the plates and the thickness of the dielectric are the same.

When the voltage across the plates is increased a point is reached where the electric field in the dielectric, V/d, reaches a value at which the dielectric changes to become a conductor of electricity. This phenomenon is known as dielectric breakdown and varies for different materials. When a conductor is placed between the plates there is no longer a capacitor, but some sort of cell depending upon the nature of the electrical conductivity.

There are instances in electrochemistry in which the "charge" stored in the systems is proportional to voltage over a limited region. For instance, if a clean nickel surface is immersed in an alkaline solution and the system allowed to equilibrate at its rest potential until the background current becomes invariant, then followed by a small step potential change applied between the nickel interface and a counter electrode, one can measure the coulombs passed until the background current is again achieved. A plot of charge versus voltage is linear up to 15 mV. In this region hydrogen is deposited upon the nickel surface and one may be tempted to use these experimental data to determine the "capacitance" from $C=q/\Delta V$.

Other electrochemical systems may undergo compositional changes that cause the rest or open circuit potential to change linearly and Conway~ presents the ruthenium oxides as being of this nature. In most electrochemical processes where solids or gases are involved the activities of the materials are invariant and the open circuit potentials indicate this by also being invariant. Sometimes there are mixed potentials due to more than one species in the product. AgO in an example in which the first product of the discharge is Ag_2O and the next step produces elemental Ag. The voltage levels for these two processes are about 0.23 volts apart. At very low current densities the difference is readily observed, but at high current densities only the lower voltage level is found. The amount of the upper plateau is rate dependent, but the voltage is not linearly dependent upon state of charge.

The relative flatness of battery discharge curves may be used in a condenser type equation for which $i*t$ is large while ΔV is small and some very high capacitance values are obtained. However, there is a hysteresis because the charge and discharge curves are not superimposible.

When solid state reactants and products are involved in electrochemical systems there are volume changes accompanying the reactions. These volume changes create stresses in the system that may lead to fatigue or even Ostwahl ripening which result in performance changes that limit utility of the system. If a system can be selected that avoids such changes useful lifetimes can be extended. One such system had alreadybe described with hydrogen atom deposition upon Ni. The H adatoms can be removed by reversing current. Other conductors can be used besides Ni including carbon. The treatment of the carbon dealing with its purity and particle size becomes important.

The particle size is not of as prime importance, however, as is the charge transfer interfacial area between the electrolyte and the electrodes. This is due to the wetting area and is analogous to the differences in BET area measurement depending upon the nature of the gas used. Again we see the importance of the interface where lectrochemistry is concerned. One side has an ionic conductor, the other side has an electronic conductor. There is an oxidation or a reduction process that accompanies the charge transfer and there is a conjugate set of occurrences at the other electrode.

In colloid science and in the field of biochemistry there is a concept called "double layer". Colloid particles and macromolecules are deemed to have a charge on their surface, and, to preserve electroneutrality, there is posited a diffuse layer of oppositely charged ions at an average distance away, δ. Thus the colloids or macromolecules are charged by ionization in some cases or by adsorption of ions from the solution. A third case was taken by analogy of frictional charge such as rubbing an amber rod with fur or sliding leather shoes across a wool rug. These latter examples are electrostatic charges and won't exist for a significant period of time in a system consisting of an insulator in a conductive solution. Ions from the solution can be attracted to sites on suspended particles that have unsatisfied or dangling bonds. Regardless of which way the double layer is created the suspended particles have one kind of charge while the solution is oppositely charged. However, these systems do not have a charge transfer interface. The surface charge does not enter an external circuit and there is no redox reaction nor is there a need for a conjugate electrode. The double layer in colloidal solutions gives rise to four related phenomena which are:

1. Electroendosmosis
2. Electrophoresis
3. Streaming Potential
4. Sedimentation Potential (Dorn Effect)

These effects are interrelated because they involved particles or macromolecules that are surface charged and, therefore, have a double layer. Electroendosmosis is the forcing of the solvent to pass through a membrane under the influence of an applied potential while the colloid is retained. Electrophoresis, conversely, forces the particle to move under an applied potential while the medium is stationary. If a dielectric solvent is forced through a membrane using pressure, a potential is observed on the two sides of the membrane and this is termed the streaming potential. If colloidal particles fall through a liquid a potential is also generated. The last phenomenon is the sedimentation potential discovered by Dorn. These four phenomena are termed electrokinetic effects and are characterized using the double layer concept. Helmholtz took the double layer concept further and postulated that the double layer behaves as a capacitor in spite of the fact that the charge source is internal and not external. When so treated he further postulated a potential that is *not* directly observable, but may be inferred by colloidal solutions gives rise to four related phenomena which are described above. The accepted name for this potential is the zeta potential. The zeta potential is a means for interrelating the four electrokinetic effects and the values of derived measurements are in agreement regardless of which of the four previously defined electrokinetic effect is used. The Helmholtz zeta potential is defined as:

$$\xi = \frac{4\pi\sigma}{D}$$

where δ is the distance between the oppositely charged sites, the charge per unit area and σ is the dielectric constant and D is the dielectric constant. Since we are dealing with ionic solutions, these are conductors and not dielectrics. The separation of charge is due to adsorption of ions or ionization of the surface as pointed out earlier.

The assumption of the dielectric constant for water, unity, is prevalent. Another difference between electrokinetic effects and electrochemical effects is that electrokinetics involves tangential motion between the plates while electrochemistry deals with phenomena occurring normal to the electrodes.

The electrokinetic effects do not involve charge passage through the external circuit and require special treatment in this book. There is yet another field that confounds electrochemistry with electrokinetic phenomena, electrocapillarity, and this must be discussed now because charge transfer is involved.

The Lippmann apparatus for electrocapillarity measurements consists of a manometer tube containing a mercury column and, importantly, a capillary tube inserted into a solution. There is an interface between the solution and the mercury column and is an electrode for our purposes. The physical location of the interface is adjusted with a mercury reservoir. There is another electrode, the necessary conjugate half-cell, a calomel electrode that is ionically connected to the Hg/electrolyte interface. The mercury column is connected to the negative end of a DC voltage source while the calomel electrode is connected to the positive terminal of the voltage source. Lippmann considers the calomel electrode to be non-polarizable while the capillary interface between the mercury and the solution is polarizable. Presumably he wants to assume no voltage change for the calomel electrode and any setting of the voltage source to be an effective change on the capillary. Potential and mercury column height are the data taken and used to determine "interracial tension" from which is derived the PZC (point of zero charge).

The above description is due to Lippmann[2,] but a revision was made by Graham[3] in 1945 in which he changed the metal from mercury to a solid, and inadvertently introduced air to the system, but otherwise left Lippmanns experimental setup unchanged. The Graham experimental setup is quite similar to another experiment in which a nickel metal immersed partially in a strongly alkaline solution was maintained at a constant potential, not by a potentiometer, but by another "electrode", the Cd, $Cd(OH)_2$/KOH system[4.] A means to detect the current flowing to the Ni electrolyte interface was included and it was demonstrated that the current was proportional to oxygen partial pressure. It was shown that the electrochemical processes were as follows:

anode $\quad Cd + 2OH^- \rightarrow Cd(OH)_2 + 2e^-$

cathode $\quad Ni + 2H_2O + 2e^- \rightarrow 2(Ni) - H + 2OH^-$

The potential versus SHE for hydrogen evolution is 19 mV more cathodic than the Cd,$Cd(OH)_2$/KOH system so that the hydrogen is adsorbed on the Ni surface/electrolyte interface. The oxygen reaching the interface reacts with the adatoms to produce water:

$O_2 + 4H - (Ni) \rightarrow 2H_2O$ a type of corrosion reaction. The interface is no longer in equilibrium with regard to the applied potential and causes the anode-cathode processes to proceed further. The rate at which the electrons travel from the Cd,$Cd(OH)_2$/KOH interface to the Ni/KOH interface is measured directly and shown to be proportional to the oxygen partial pressure. Because the Cd,$Cd(OH)_2$/KOH electrode is a battery electrode the voltage change with current is small and is really quite independent of the state of charge, which is the quantity of active material remaining, providing the measurements are made near open circuit and sufficient time is allowed to reach a stationary state.

The electrocapillarity measurements ignore the current passage and the effect of atmospheric gases. Even modern measurements relying on very thin gold electrodes and the quartz microbalance also ignore the charge transfer processes at the interface. There is a question of buoyancy of thin electrodes by deposition of gases and the experiment should also include coulometry.

The conclusion drawn from the above discussion is that the double layer capacitor does not send charge through the external circuit.

Operation of these devices is not as DLCs or "Ultracapacitors", but rather as Supercapacitors. Seemingly capacitative behavior is observed with some faradaic processes. Two examples would be the case of adatoms such as deposition of a layer of hydrogen on elemental nickel, and when there is a "fish tail" during battery discharge due to more than one entity as in the nickel oxide electrode. Capacitative behavior is defined as voltage proportional to coulombic charge. This behavior is known as pseudocapacitance to differentiate it from the double layer concept. There is a major difference between the pseudocapacitance and the conceptual double layer capacitance which is simply that one is faradaic in nature and the other is not. For this reason their behaviors and one can not consider the two to affect one another as do two electrostatic capacitors.

Another difference between electrostatic capacitance and faradaic pseudocapacitance is the so-called electrochemical series resistance (ESR) which is just another way of indicating hysteresis in the charge-discharge curves. There are significantly different voltages at the same current density depending upon energy storage or energy removal.

The major differences between energy storage in a battery and in a supercapacitor are that the battery has a major volume change for the active materials as opposed to virtually no volume change in the supercapacitor, and also the battery electrode/electrolyte interfaces have little change in state of charge whereas the supercapacitor electrodes/electrolyte interfaces have large changes in state of charge since all charge is stored in a stationary plane. The battery has a moving reaction zone. It is the difference that Conway refers to as two dimensional and three dimensional processes.

1   B. Conway, J. Electrochem. Soc., *138, 1539 (1991)*.
2   G. Lippmann, Pogg. Ann. *149*, 546 (1878), Ann. Chim. Phys. 5, 494 (1875), *12*, 265 (1877).
3   D.C. Grahame, J. Chem. Phys. *18*, 903 (1950); Chem. Rev. *41*, 441 (1947).
4   H.N. Seiger, papers presented at the ECS Meeting, October 1964, New York.

V

Membrane Potentials

Membrane potentials arise during consideration of the Gibbs-Donnan[1,2] membrane phenomenon. This effect is important is physical biochemistry as well as in colloid science and is worthy of our consideration. There are colloids or macromolecules that are ionizable. If a membrane separates the macromolecules and a salt solution having an ion in common with the macromolecule the salt can diffuse but the macromolecule is specifically selected so as to be impermeable. One such material is the dye congo red which dissociates into Na^+ ions and the ionic dyestuff represented by R^- Gelatin chloride solutions have also been used in Donnan Effect studies.

Conceptually set up an experiment in which the colloid is in an inside compartment separated from the outside compartment by a membrane. The outside compartment contains salt at concentration C_2. The colloid is at concentration C_1. The initial condition is represented by the following:

Na^+	R^-	Na^+	Cl^-
C_1	Cl	C_2	C_2
inside		outside	

and after permitting equilibrium to ensue some NaCl leaves the outside compartment and enters the inside compartment which is represented by the following conditions:

Na+	R-	Cl-	Na+	Cl-
Cl + x	Cl	x	C2 - x	C2 - x
	inside		outside	

Electrical neutrality is maintained by this distribution and the Gibbs postulate is that the product of the diffusible ions on both sides are equal, i.e.:

$$(C_1 + x)x = (C_2 - x)^2 \text{ , which simplifies to :}$$

$$x = \frac{C_2^2}{C_1 + C_2}$$

Suppose the starting macromolecule was the sodium salt of congo red at a concentration of 2mM inside the membrane. The NaCl concentration outside the membrane is initially set to 10 mM. These values are C_1 and C_2 respectively, and solving for the concentration changes x = 0.0045. The two compartment conditions are now represented by:

Na+	R-	Cl-	Na+	Cl-
0.0065	0.002	0.0045	0.0055	0.0055
	inside		outside	

This distribution represents a Donnan Equilibrium in which the outside compartment has decreased concentration by 45%. If measurements are made with appropriate reference electrodes, we recognize the concentration cell for which the voltage difference is

$$E = \frac{RT}{F} \ln \frac{0.0055}{0.0045}$$

calculating to 5.1 mV. Measurements for a gelatin chloride system as the salt of the impermeable ion of gelatin along with HCl in the outer compartment had been made. The HCl concentration was measured by pH determinations and membrane potentials were calculated from these pH values and compared to direct measurements using saturated calomel electrodes. Historically, other "electrodes" were used, but our treatment of the subject shall be more readily understood using the calomel electrode treatment first and then proceed to the other electrodes.

It is important to treat this system theoretically to gain an insight into the so-called direct measurement of membrane potential. First let us recognize that pH measurements are of a salt or acid solution and not just the hydrogen ion. Thus, the solution is HCl in water and not just H^+ ions. Now, look at the experiment as carried out by Loeb[3]:

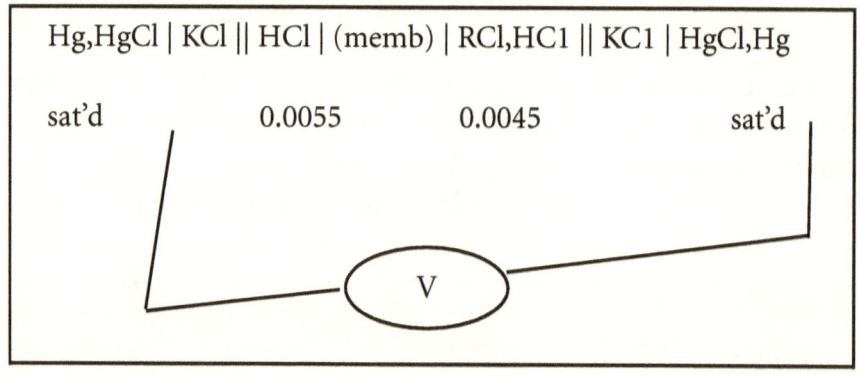

$$Hg,HgCl \mid KCl \parallel HCl \mid (memb) \mid RCl,HCl \parallel KCl \mid HgCl,Hg$$

sat'd   0.0055   0.0045   sat'd

V

outside   inside

This is recognized as a concentration cell since the concentration of HCl differs on both sides of the membrane and yields the 5.1 mV calculated above. In the outside compartment the concentration of HCl is greater than the inside compartment and the tendency is for the electrochemical processes to decrease the HCl concentration. The chloride ion migrates across the solid/liquid interface to combine with a mercury atom that, in turn, releases an electron to the outer circuit:

$$Cl_{C2}^- + Hg \rightarrow HgCl + e^-$$ which makes the more concentrated, outside compartment, negative with respect to the other. The current

carrier in the solution is the proton to which the membrane is permeable. The counter electrode has the reduction process, namely:

$$e^- + HgCl \rightarrow Hg + Cl^-$$ resulting in the increase of Cl^- ion, i.e. HCl, in the outside compartment. In principle this process continues until the HCl activities on both sides are equal, but using electronic digital meters, the current drain is negligible and the process of equalization of pH occurs very slowly. It is of interest to note that the potential has been ascribed to the membrane. In particular Loeb has data which show the membrane potential varying from 2 to 30 mV depending on pH and concentrations.

Other electrode systems had been used historically in measurement of membrane potentials. A popular one has been the standard hydrogen electrode. The standard hydrogen electrode (SHE) is a platinum electrode partly immersed in an acidic solution of unit activity which has pure hydrogen of unit fugacity bubbling. The acidic solution is thereby saturated with hydrogen. A pair of such electrodes are introduced, one to the inside compartment and the other to the outside compartment via salt bridges. Potential readings are taken and are found to be zero. Since independent measurements of the solutions on both sides are also taken, some value is calculated and ascribed to the system. Because the measurement and the "expected" values differ, it is inferred that a "membrane potential arises which is equal and opposite to the 'expected'" result. It is by such reasoning that membrane potentials are defined. It is noted, now, that the "membrane potential" is an artifact of the method used to create its definition. Let us investigate how this comes about by comparing measurements made with a silver, silver chloride electrode which is similar to the calomel electrode used by Loeb and with SHE's.

Consider the system used by Loeb consisting of gelatin chloride and HCl. This is diagrammed as follows:

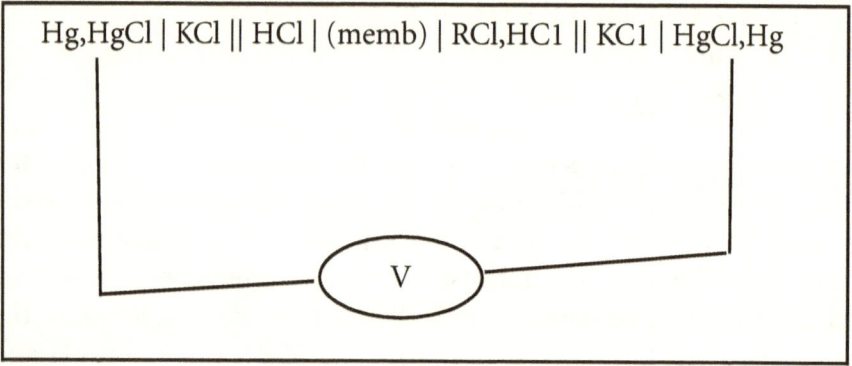

Hg,HgCl | KCl || HCl | (memb) | RCl,HCl || KC1 | HgCl,Hg

The initial conditions are:

outside: $H^+(C_2)$ $Cl^-(C_2)$
inside: $R^+(C_1)$ $Cl^-(C_1)$

After 20 hours the conditions become:

outside: $H^+(C_2-x)$ $Cl^-(C_2-x)$
inside: $R^+(C_1)$ Cl^- (C_1+x) $H^+(x)$

Electrical neutrality is preserved even with the changes due to the Donnan Effect. As a result the HCl concentration in the outside compartment does not equal the HCl concentration of the inside compartment. When the circuit is completed using the meter labeled V, the HCl concentration inside wants to decrease. To accomplish this silver must react with the Cl^- ion:

$$Ag + Cl^- \rightarrow AgCl + e^-$$ and the H^+ ion must leave the outside compartment and migrate through the membrane to the inside compartment. The outside compartment is negative with respect to the inside compartment. The concentration changes occur because Ag and AgCl (Hg and HgCl in case of the calomel electrode) are both insoluble and are electroactive. The potential difference measured depends upon the difference in HCl concentration in the two compartments. This explains the results of Loebs[3] experiments using calomel electrodes.

When SHE's are used there are differences because the acid concentrations in the two compartments have no mechanism for changing with the passage of current. Diagramatically the experiment is as follows:

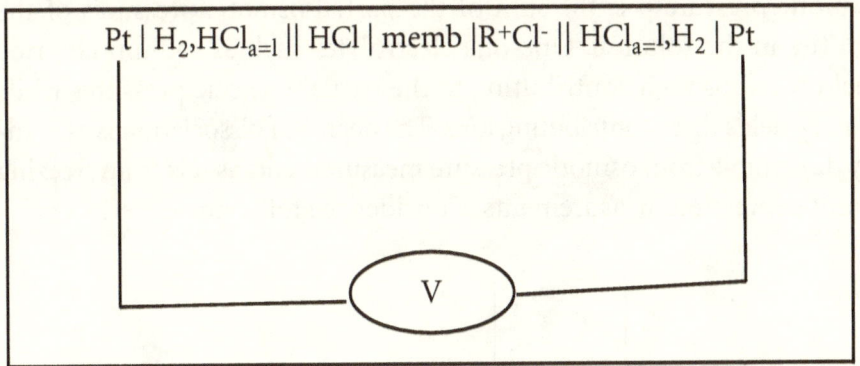

$$\text{Pt} \mid H_2, HCl_{a=1} \mid\mid HCl \mid \text{memb} \mid R^+Cl^- \mid\mid HCl_{a=1}, H_2 \mid \text{Pt}$$

V

Pt does not react with Cl^- so its concentration remains invariant. H_2 gas also does not react with Cl^- ion. With no driving force there is no change; the free energy is zero and the voltage measurement is also, consequently, zero.

Osmosis is found to follow the van't Hoff equation which is analogous to the ideal gas law:

Gas Law: $PV = nRT$

van't Hoff
$$\pi = \frac{C}{MW} RT$$

In the ideal gas law n is the number of moles of gas and V is the volume occupied by the gas. If the volume is given in liters then n/V is a concentration term in moles per liter. Analogously, the van't Hoff relation has the solute present in the solvent in grams per liter of solution and MW is the molecular weight so that C/MW is also moles per liter, molarity.

In a gaseous system equilibrium is encountered when pressure and temperature are everywhere the same. Using the van't Hoff equation/ideal gas law analogy we shall assume that the liquid on the two sides of a semipermeable membrane are in equilibrium when the temperature and osmotic pressures are everywhere the same. As the total gaseous pressure is the sum of the partial pressures of the gases, the total osmotic pressure p is the sum of the partial osmotic pressures of the entities in the solution. One-one electrolytes such as sodium chloride yield two ions each contributing to the partial osmotic pressures while $CaCl_2$ yields three contributing ions. The degree of dissociation is as readily discernible from osmotic pressure measurements as it is from freezing point depression measurements. Consider the following experiment:

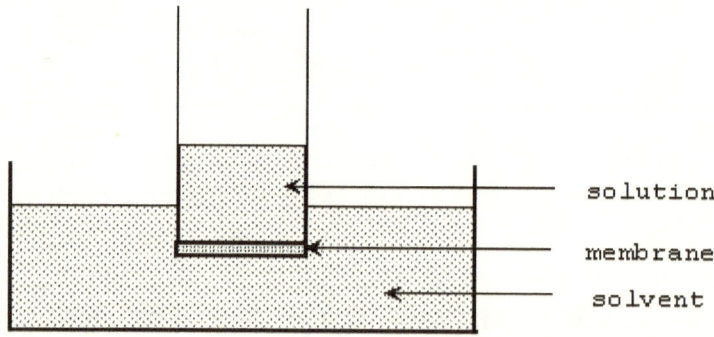

The membrane separating the solution from the solvent is semipermeable because the solvent is permeable through the membrane while the solute is not. At the start of the experiment the solvent levels and the solution levels are the same. At equilibrium sufficient solvent permeated to the solution compartment to raise its level by height h. The pressure on the solvent is $A + v^o$ where A is the barometric pressure and v^o is the vapor pressure of the pure solvent. The pressure on the solution is $A + v' + hdg/area$ where v' is the vapor pressure of the solution and the last term is the weight of the solution (column height h, solution density d and g the gravity constant). Now apply the equilibrium conditions: $A + v^o = A + v' + hdg/area$ which induces us to neglect the barometric pressure. Note that if the cylinder in which the solution rises has a small diameter the solution height has to be corrected for surface tension. We shall neglect this correction. Rewriting the equation as:

$$v^o - v' = hdg/area = \pi$$ and using Raoult's Law:

$$v' = v^o N_1$$, where N_1 is the mole fraction of the solvent, then:

$$v^o - v^o N_1 = \pi = (1 - N_1)v^o$$ but $1 - N_1 = N_2$, yielding

$$\pi = v^o N_2$$, so that the osmotic pressure is equal to the product of the vapor pressure of the solvent multiplied by a concentration term for the solutes. Putting in an appropriate conversion term for mole fraction to any other concentration such as molarity, then:

$$\pi = v^o kM$$

In the van't Hoff form $v^o k = RT$ and M are still molarity.

We are now prepared to look at the Donnan Effect as an osmotic pressure problem, and use Loebs gelatin chloride work as a specific example.

G+	Cl-	H+	H+	Cl-
M_1	$M_1 + x$	x	$M_2 - x$	$M_2 - x$
	inside		outside	

1. Inside Initial Condition: $\pi_G + \pi_{cl} = \Pi_{inside}$, but the chloride salt of gelatin is ionized yielding two entities and $\pi_G = \pi_{cl}$ since $\pi_G = M_1 RT$ and $\pi_{Cl} = M_1 RT$ RT so that:

$$\Pi_{inside} = 2M_1 RT$$

2. Outside Initial Condition: $\pi_{H^+} + \pi_{Cl^-} = \Pi_{outside}$, similarly, the HCl is completely ionized and $M_{H^+} = M_{Cl^-}$ so that:

$\Pi_{outside} = 2M_2RT$. Let the solvent and solution reach equilibrium with respect to the temperature and osmotic pressure so the equilibrium value is:

$\Pi_{outside} = \Pi_{inside}$ and

$\Pi_{inside} = \pi_G + \pi_{(M1+x)}$

$\Pi_{outside} = 2\pi_{(M2-x)}$,or

$\pi_G + \pi_{(M1+x)} + \pi_x = 2\pi_{(M2-x)}$

Using the van't Hoff form:

$M_1RT + M_1RT + xRT + xRT = 2M_1R - 2xRT$, simplifying:

$2M_1 + 2x = 2M_2 - 2x$

$M_1 + x = M_2 - x$

$X_S = \dfrac{M_2 - M_1}{2}$

The last equation is for the molar loss of HCl from the outside compartment that enters the inside compartment to achieve an equality of osmotic pressure. The Donnan Equation given previously is rewritten here for convenience as:

$X_D = \dfrac{M_2^2}{M_1 + M_2}$

The two equations are certainly quite different, but taken to the limit as the outside diffusible material becomes more concentrated, both equations approach the same value, $M_2/2$. This leads to the quantitative question of the range of concentrations encountered experimentally.

Loeb used gelatin chloride at a level of 1 gram per liter of water inside the membrane and pH between 2 and 4 outside the membrane. Gortner reports that one gram of gelatin combines with 1.7×10^{-4} equivalents of Cl^- ion. Now, if there is one replaceable anion site per molecule, the molecular weight of gelatin is at least 5880 grams. The concentration of the gelatin is 1 to 2 mM. The acid starts also in the range of 1 mM and increases. Donnan ran his calculations up to 100 times greater in one compartment than in the other. We can quantitatively look at the two using the van't Hoff form of equations by setting $M_1 = 1$ mM and letting M range from values of 1 mM to 100 mM and solving for in the two cases. The calculations were done using Mathcad Version 4.0 for Windows making the following substitutions: for x_s use x_a and for x_D use y_a. Let $M^1 = 1$ mM and $M_2 = a$. The range of a is between 2 mM and 100 mM. The results are shown in Figure V-1. The plot of the expressions show that in the experimental region the two curves are but little different. This result suggests that the Donnan Effect can be explained using osmotic considerations.

The newly derived equation for the Donnan Effect becomes even more informative when combined with the Nernst Equation. Let the initial concentrations of the impermeable ion be M_1 and of the permeable one-one electrolyte be M_2. The Donnan Effect equation in these terms is:

$$X_s = \frac{M_2 - M_1}{2}$$

and the Nernst equation for the equilibrium condition is:

$$V = \frac{RT}{zF} \ln \frac{M_2 - x}{x}$$

simplifying to:

$$V = 0.059 \log \frac{M_2 + M_1}{M_2 - M_1}$$

The simplified form is used to expect that the potential difference between the two compartments decreases toward zero as M_2 becomes larger with respect to M_1. This is what is expected as the osmotic pressures in the outside compartment increases and more salt enters the inside compartment. The equation also contains a singularity about $M_1 = M_2$ and is not useful except in the region where $M_2 \rangle M_1$. When $M_2 \langle M_1$ the Donnan Phenomenon does not take place, but rather the driving force towards equilibrium is to have the permeable ionic species to move from inside to outside the membrane. This obverse phenomenon is recognized as dialysis.

Treating membrane phenomena on the basis of the van't Hoff equation for osmosis and employing electrochemistry with implicit charge transfer the following areas are understood:

Membrane potentials
Donnan Effects
Dialysis

This treatment gives rise to a self-consistent theory that explained the so-called "membrane potentials", the Donnan Effects and causes dialysis to be just a special case of membrane phenomena that occurs when the osmotic pressure within the membrane compartment is greater than that of the outside compartment.

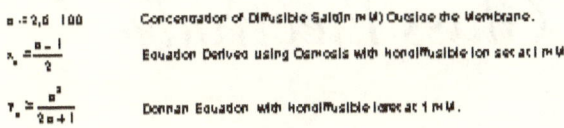

Comparison of the Gibbs-Donnan Equation with That Based on Os Equilibrium In the Concentration Range Typical of The Donnan Phenomenon

$a = 2, 6 \ldots 100$ Concentration of Diffusible Salt(in mM) Outside the Membrane.

$x_a = \dfrac{a-1}{2}$ Equation Derived using Osmosis with Nondiffusible Ion set at 1 mM.

$y_a = \dfrac{a^2}{2a+1}$ Donnan Equation with Nondiffusible Ions at 1 mM.

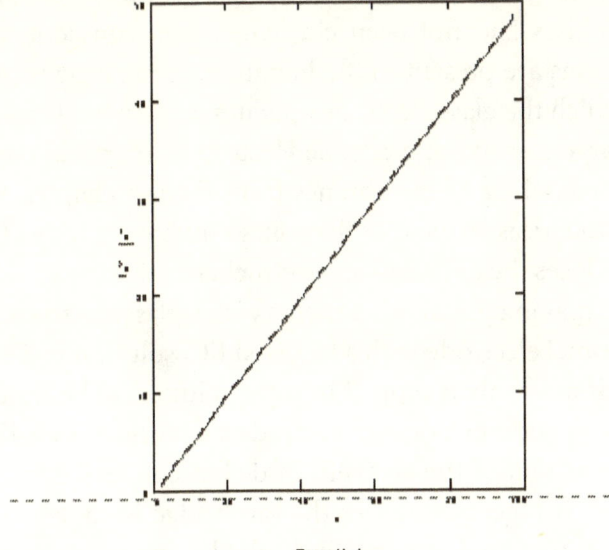

Figure V -1

1 W.Gibbs, *Collected Works*, Longmans, Green, N.Y., 1928.

2 F.G. Donnan, Z. Elecktrochem. 17, 572, (1911).

3 J. Loeb, *Proteins and the Theory of Colloid Behavior*, McGraw Hill, 2nd Ed. New York, 1924.

4 R.A. Gortner, *Outlines of Biochemistry*, J. Wiley & Sons, New York, 1949.

VI

The Glass Electrode

The glass electrode is very commonly used for precise pH measurements. The function of the glass had not been elucidated and, consequently, divergent interpretations are present in the literature. Some insight into the mechanism by which the glass electrode operates may be made using the principles of charge transfer at electrode/electrolyte interfaces combined with the understanding of membranes from the last chapter. We must not forget that the measurement of the voltage difference across the terminals effectively closes the circuit of an electrochemical cell.

The procedures in making pH measurements with a glass electrode is, usually, to have a calomel electrode with a buffered KCl solution inside a glass membrane. This assembly is dipped into the solution to be tested. The second electrode is another calomel electrode surrounded by a KCl of the same molarity as that of the glass electrode but connected to the test solution via a salt bridge. Sometimes the salt bridge is an asbestos fiber, or a glass frit or an actual inverted "U" tube having a salt bridge that may be liquid or gelled. Expressing this information as an electrochemical cell, which it is, we may write:

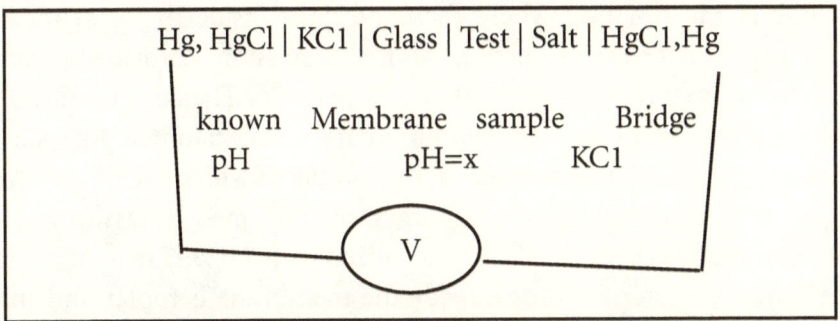

First we want to identify what material is oxidized and what material isreduced when current is passed to identify the charge transfer processes. It is not the potassium ions, nor the chloride ions. This leaves the mercury to be oxidized in the anode compartment and the HgCl to be reduced in the cathode compartment. Further, we have the requirement that only if there is a force will there be a potential. That force is the difference in activity of the acids on either side of the glass membrane. Thus, there is a migration of protons across the membrane to tend to equalize the activities on the two sides of the membrane. It is assumed that the test sample has a higher value of pH than the glass electrode so that protons flow from the glass electrode compartment to the sample compartment. Electronic charge flows through the external circuit which is the potential measuring device, V, in the diagram. The anode reaction is:

$$Hg + Cl^- \rightarrow HgCl + e^-,$$

where the Cl^- comes from the solution and the electron enters the external circuit. The cathode reaction is simply the reverse:

$$e^- + HgCl \rightarrow Cl^- + Hg$$

Conduction through the membrane is by migration of protons, but diffusion processes account for conduction in the electrolyte of the compartments.

Look at the situation where the conditions inside the glass membrane consists of mercury metal, calomel, potassium chloride at some concentration and an acetate buffer at a pH 4.76. The conditions outside the glass membrane are, again mercury metal, calomel, potassium chloride at the same concentration, the sample of which the pH is to be determined is Clarke and Lubs buffer of 0.05 meq potassium acid phthalate that has been partially neutralized with 0.02995 meq of KOH. The concentration of protons inside the membrane is molar and that on the other side is molar, resulting in a driving force for transfer of protons through the glass to the external sample. Protons leaving the membrane represent a decrease in acidity while the increase in the sample compartment represents an increase moving towards an equilibrium. The rate of change is the current flowing through the external circuit which accounts for electrical neutrality everywhere within the system. However, by using very high impedance circuitry the rate of diffusion is extremely small so that readings have imperceptible differences during the time required for determination of the potential differences.

As pointed out earlier, because the concentration of protons within the glass electrode is greater than that of the other compartment, the glass electrode is negative with respect to the other electrode so that the electrons flow from this terminal to the other terminal while protons with their positive charge migrate from one compartment to the other through the glass membrane. Because of the requirement of electrical neutrality the changing acid quantity on both sides may be represented

as $HgCl$. The $HgCl$ concentration inside is decreased because protons move across the membrane while chloride ions support the newly

Hg^+ oxidized ions by precipitation. Similarly Cl^- ions are formed on the other side at a rate equal to the arrival of protons. Thus, the glass electrode behaves as a concentration cell and we may readily write the equation to relate the potential difference between the two compartments to the concentrations or activities of HCl in the two compartments:

$$\Delta V = \frac{RT}{zF} \ln \frac{(HCl)_2}{(HCl)_1} = 0.059 \times \log \frac{(HCl)_2}{(HCl)_1}.$$

For the concentrations with which we are dealing AV is 44 mV.

Thus, we have elucidated the mechanism by which a glass electrode operates by recognizing the concentration cell aspect. Furthermore, pH can be defined in terms of the activity of strong acids which is measurable by such means as boiling point elevation or freezing point depression instead of hypothesizing about a mean activity coefficient of a single ion. we keep our faith with Guggenheim.

VII

Hittorf Transference and Electrolytic Phenomena

In 1853 Hittorf postulated that the current between two electrodes immersed in an electrolytic solution was carried by the ions in the solution. The cations and anions both migrate but in opposite directions and the rates of these migrations are not necessarily the same. The fraction of the total current carried by the cations is termed the cation transference number, t_+, and the anion transference number, t_- is that fraction carried by the anions. The sum of t_+ and t_- is unity so that any experiment to determine one transference number results is determination of the other as well.

From the concept of transference, ionic mobility had been derived. Mobility has been defined as the velocity at which an ion travels in a potential gradient of one volt per centimeter. Application of a potential across the electrolyte is not a simple process. Applying such a field necessitates (1) the external source and (2) a pair of conjugate electrodes. Since an oxidative process occurs at the anode and a reductive process occurs at the cathode, the applied potential drops at the interface so that any measured potential across the electrolyte is the residual.

Suppose a cell is made for the measurement of mobility as shown in Figure VII-1. This figure is historical and discussed by Glasstone[1.] It has been updated to include an ammeter and a coulometer. This cell is recognized as a battery in which the cathode is $AgCl$ and the anode is Cd. Writing the electrode processes to represent the conjugate reactions we obtain:

48

$$2AgCl + 2e^- \rightarrow 2Ag + 2Cl^-$$

$$Cd + 2Cl^- \rightarrow CdCl_2 + 2e^-$$

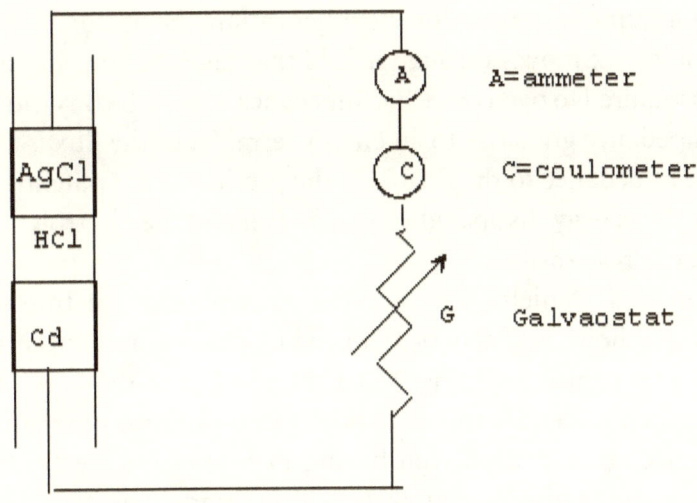

Figure VII-1.

but $CdCl_2$ is soluble in water so that it is ionized to Cd^{++} and $2Cl^-$, and summed the processes become:

$2AgCl + Cd \rightarrow 2Ag + Cd^{++} + 2Cl^-$. The ions in the electrolyte are H^+, Cl^- and Cd^{++}. if one looks at the anode and cathode processes and knows they are electronically coupled the rates are identical. Also, by the Kirchhoff current law the Cl^- flux into the anode and the Cl^- flux from the cathode and every section in the electrolyte has the same flux, it can be unequivocally stated that the current between the electrodes is represented by the chloride ion flux and not by Cd^{++} ion diffusion nor proton diffusion. Inequalities represented by the increasing Cd^{++} ion concentration are joined by an equivalent Cl-ion concentration to maintain electrical neutrality everywhere in the solution. That is to say, any local diffusive motion of Cd^{++} carries along two Cl$^-$ ions as an ion pair and does not contribute to the current flux between the electrodes. A similar argument is made for the other cation, the proton.

The ionic flux moves through a field that tends to impede the flux. Energy is required to overcome this impedance so that the flux multiplied by the impedance gives rise to an energy term. Since the flux is current, I, and the impedance to the flux is, R, the product is the potential drop, V. This is the energy dissipated inside the cell and the electrolyte resistance is partly responsible.

Getman and Daniels[2] have another example for the transference numbers and how they may be deduced by changes in concentration. Their figure is copied here (Fig. 113, p 402) as Figure VII-2. While they do not explicitly identify the electrode materials, from time and place we recognize an electrolysis cell having Pt immersion electrodes. For electrolysis the cathode compartment is made negative and the cathodic part of the process is:

$$2H^+ + 2e^- \rightarrow H_2$$

while the anodic part is:

$$2Cl^- + H_2O \rightarrow \frac{1}{2}O_2 + HCl + 2e^-$$

Note that Cl$^-$ is not oxidized since the redox potential for oxygen is lower than that for chlorine. Again, in equivalents the oxygen and hydrogen evolution rates are equal and so is the H$^+$ ion flux. Again, by Kirchhoffs current law only H$^+$ ion is transported.

A third case for transference number deals with electrodes that participate in the reactions at the electrodes. This is a copper anode, platinum cathode with a $CuSO_4$ aqueous electrolyte. In this case Cu^{++} ions are generated in the anode compartment, Cu^{++} ions are plated from the cathode compartment:

anode: $Cu \rightarrow Cu^{++} + 2e^{-}$

cathode: $Cu^{++} + 2e^{-} \rightarrow Cu$.

Similar arguments based on conjugate electrodes and Kirchoffs current law indicate no changes in concentration of the Cu^{++} ion in any compartment.

These considerations fly into the face of what has been taught for transference and also fly into the face of many, many experiments for the Cu/CuSO$_4$/Cu transference cell which is the experiment performed by undergraduates in physical chemistry laboratories.

In order to achieve concentration changes in the anode or cathode compartments, one has to propose a parasitic or competitive reaction. Hence, we deal with a process efficiency. In the case of the Cu anode immersed in a CuSO^4 solution we shall treat the instance of corrosion of a partial fuel cell reaction competing with Cu anodization, and at the cathode a codeposition of Cu and H_2 so that the Cu is contaminated with the hydrogen. These competitive reactions alter the electrode process efficiencies at both electrodes. Let us write the reactions that occur:

anode $Cu \rightarrow Cu^{++} + 2e^{-}$

cathode (main) $Cu^{++} + 2e^{-} \rightarrow Cu$

(competitive) $2H_2O + 2e^{-} \rightarrow H_2 + 2OH^{-}$

Net $Cu + 2H_2O \rightarrow Cu(OH)_2 + H_2$

This competitive reaction alters the electrode process.

From the set of reactions it becomes obvious that changes in concentration of Cu content in the terminal compartments are consistent with electrode process inefficiencies. These concentration changes can be fitted into the Hittorf relationship to calculate a "transport number."

1   S. Glasstone, *Textbook of Physical Chemistry*, Second Edition, D. Van Nostrand, New York, 1946.

2   F.H. Getman & F. Daniels, *Outlines of Physical Chemistry*, J. Wiley & Sons, New York, 1947.

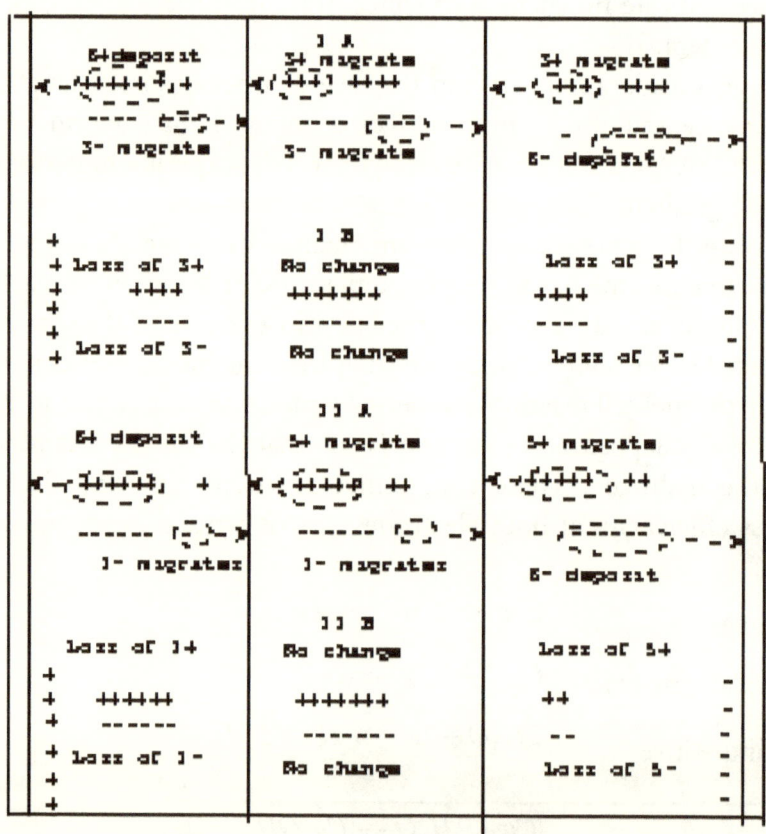

Figure 1-1. Getman and Daniels Migration Diagram.

VIII

Liquid Junction Potentials

A potential at the interface between two different electrolytes had been postulated. One example is to have solutions of HC1 in similar cells, one having a concentration of m_1 molal and the other half-cell electrolyte at a concentration of m_2. The potential is presumed to arise at the interface of the two liquids. To overcome the potentials salt bridges are proposed. The salt bridges consist of solutions of KC1 selected because the "transference numbers" of the K^+ ion and the Cl^- ions are nearly equal. Other means of avoiding the junction potential consist of using agar bridges with KC1 in the agar, sintered glass frits and asbestos threads. In each of these there is ionic conductivity so that the system is an electrochemical cell when some electronic connection is made externally.

Figure VIII-1.

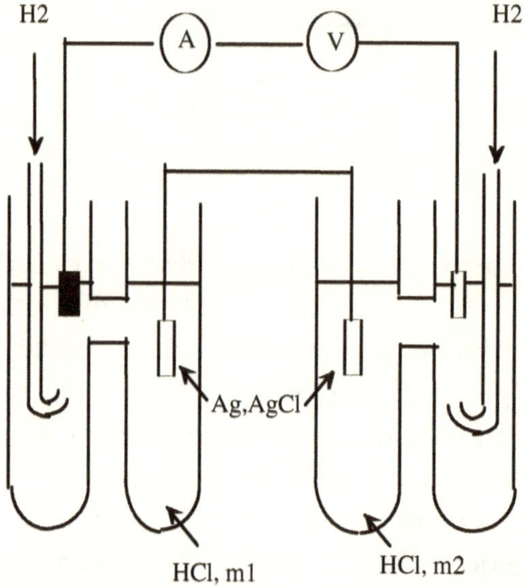

A = ammeter

V = voltmeter

For the purposes of our discussion we shall make sure the cell voltages are in opposition so that only a differential voltage is measured. The experimental circuit is shown in Figure VIII-1. These are two silver chloride/hydrogen battery cells that are similar except for the concentration of electrolyte, an aqueous solution of HCl of concentration m_1 in the left hand cell and m_2 in the right hand cell. In the left hand cell the hydrogen electrode consumes H_2 at the rate read by the ammeter, A:

$$H_2 \rightarrow 2H^+ + 2e^-,$$

with the reduction reaction in the left hand cathode coupled to it:

$$2AgCl + 2e^- \rightarrow 2Ag + 2Cl^-,$$

for an overall reaction:

$$H_2 + 2AgCl \rightarrow 2Ag + 2HCl$$

which is responsible for an increase of the HC1 concentration in the left hand cell.

To obtain the voltage across the left hand cell we may write:

$$K = \frac{(Ag)^2 (HCl)_l^2}{(H_2)(AgCl)^2}$$

$$\Delta G = \ln K = -zFE_l$$

$$E_l = -\frac{RT}{zF} \ln \frac{(HCl)_l^2}{(H_2)}$$

For the right hand cell the potential may be written as:

$$E_r = -0.030 \log \left(a_{HCl} \right)_r$$

The reaction rate is governed by the value of ΔE and the load impedance. The load impedance is the sum of the effective impedance of the two external meters and the internal impedance. If the experiment is arranged so that the effective external resistance is very high, say in the order where current may be neglected, the reading of the voltmeter, V, is ΔE. Then:

$$V = E_l - E_r$$

$$V = \left[-0.030 \log \left(a_{HCl} \right)_l \right] - \left[-0.030 \log \left(a_{HCl} \right)_r \right]$$

$$V = -0.030 \log \frac{\left(a_{HCl} \right)_l}{\left(a_{HCl} \right)_r}$$

which is a concentration cell voltage. For a 10 fold difference in concentration between the two compartments, say 1 molar and 0.1 molar the voltage difference is 30 mV.

Now, consider the second case on page 465 in Getman and Daniel's textbook shown here as Figure VIII-2. There are two ways to equalize the concentration differences of the two compartments. The first way is by diffusion of LiCl from the more concentrated compartment to the less concentrated compartment. The second is by an electrochemical means in which an external electronic connection is made. This occurs whenever a potential reading device is placed across the battery. Several means for making the diffusion rate small are available, but no matter what method is chosen, the separation between the two compartments must permit ionic conductivity or else no potential differences could be detected and measured.

Figure VIII-2.. Liquid Junction Measuring Cell (per Getman and Daniels)

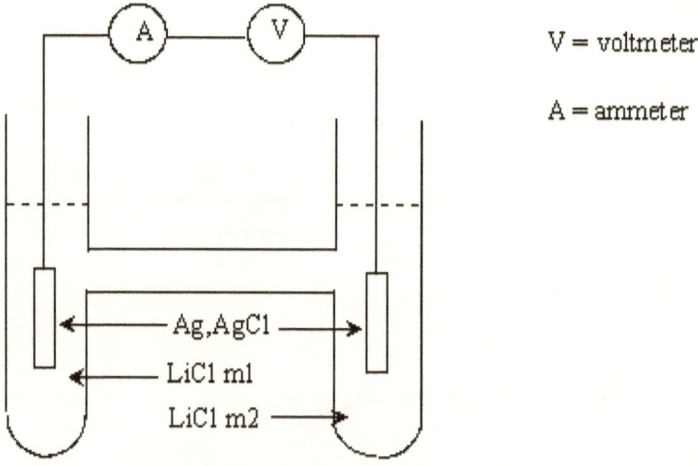

V = voltmeter

A = ammeter

Under these conditions the reaction in the more concentrated electrolyte side becomes:

$$Ag + Cl^-_{a1} \rightarrow AgCl + e^-$$

with the reaction in the less concentrated compartment being:

$$AgCl + e^- \rightarrow Ag + CL^-_{a2}.$$

By this electrochemical means the Cl⁻ ion concentration in the more concentrated compartment is decreased while it is increased in the other. This is not the whole answer. The ionic conductivity is supplied by migration of Li^+ ions to fulfill the requirements of electroneutrality. This experiment can be set up in two different ways. In one there is an inhibition to Li^+ ion migration so that when the measurement is made the electrolyte concentrations are unchanged. In the second case the impedance to Li^+ ion migrations is small and very quickly the activities of LiCl in the two compartments are equilibrated, and there is a relatively high current which also gives rise to polarizations so that V quickly approaches zero. If we now identify the case of inhibition to Li^+ migration as a cell with no "transference", the voltage will be as previously calculated:

$$V = 0.03\log\frac{a_2}{a_1},$$

and the one with "transference" the voltage approaches zero as the concentrations of the two compartments equilibrate.

IX

Electrophoresis

There is an interesting experiment performed by carrying out electrolysis in a 2.5 cm diameter Tygon tube. The set up is shown in Figure IX-1:

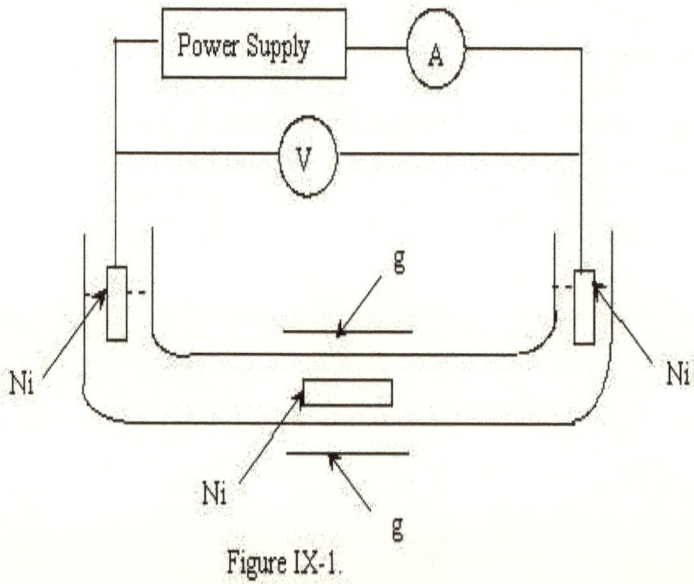

Figure IX-1.

A nickel bar 3 mm by 12 mm by 55 mm was placed in the tube as shown. A vice grip, g, with 75 by 25 mm jaws were placed about the nickel bar in such a way that the tube could be flattened at will. The Tygon is in the shape of a "U" and two Ni plates are inserted as shown in the sketch. The tube was filled with 6 M KOH solution to partially cover the terminal Ni plates.

The power supply was able to maintain a constant current which an operator may select. A current was chosen so that bubbles were apparent on the two Ni plates. At the one made negative H gas was evolved:

$2K^+ + 2H_2O + 2e^- \rightarrow 2KOH + H_2$, and at the positive electrode there are two reactions competing:

$2OH^- \rightarrow H_2O + \frac{1}{2}O_2 + 2e^-$, and

$Ni + 2OH^- \rightarrow Ni(OH)_2 + 2e^-$

While letting these reactions proceed at a constant rate, the vice grips, g, were slowly closed. As the tygon tubing shape was changing, the dimensions of the path around the Ni bar changed to as to increase the impedance in this region and the voltage across the terminals rises to maintain the constant current. The increase is, mainly, the drop along the nickel bar. When this drop reached a high enough value, $\Delta V \approx 1.5$ AV volts , bubbles arose from the two ends of the bar which, from volumes of gas emitted, are O_2 and H_2. There may still be still be an electrolyte path in parallel with the Ni bar but the current and impedance have a product, voltage drop, large enough for the electrodes to be formed on the two ends of the bar.

The potential gradient in the electrolytic solution is not the total voltage applied to it divided by the path length, but rather is the impedance of the electrolyte multiplied by the current. Direct measurements can be made using a pair of reversible electrodes, say calomel or

Ag / Ag_2O or $Ag / AgCl$, contacting the solution with a fixed distance between them. To avoid errors the change in voltage between the measuring electrodes should be 0.1 to 0.2 volts, and the readings should be done very quickly to avoid polarization of the measuring electrodes.

The above experiment allows us to understand the phenomenon known as electrophoresis. Colloidal solutions, say a mixture of proteins, are placed into an electrolytic solution such as a dilute KC1 solution and a high voltage applied. This applied voltage is between two terminal electrodes which are poised, say $Ag / AgCl$, with the two silver species in equivalent amounts. The current/voltages chosen by design are of such a magnitude that the electrode reactions are substantially

$Ag + Cl^- \rightarrow AgCl + e^-$ for the anode and $AgCl + e^- \rightarrow Ag + Cl^-$ for the cathode. Such design is accomplished by current density considerations for the electrodes and the effective impedance of ~he solution undergoing electrophoresis.

If the cross section of the electrophoretic device is uniform then

the colloids are in a potential field of $L \times j$ where L is the effective impedance per unit area and j is the current density. The colloids then migrate in the potential field according to their charge and molecular size and shape. Thus the colloids are separated from one another during electrophoresis. There is a rough similarity between electrophoresis and time-of-flight mass spectrometry. The subject of application of a magnetic field to an electrophoretic device shall not be treated and left as an open question.

X

Natural Convection in Electrochemical Systems

Electrochemical reactions in which ionic diffusion is rate determining become more rapid with stirring. The rates are increased because the thickness of the diffusion layers is decreased resulting in a greater diffusive flux. The rotating disk electrode is a good example; it is characterized by having a uniform thickness of the diffusion layer. Any stirring by a propeller, magnetic bar or gas flowing decreases the average thickness of the diffusion layer. These, however are all examples of forced convection and we shall concern ourselves with natural convection, sometimes called free convection by hydrodynamacists[1].

Natural convection arises when density gradients are favorable. Simple examples and usefulness of the concept are seen in the frozen food section of supermarkets where upright freezers have doors and horizontal freezers are frequently uncovered. The density of the cold air is greater than the ambient air so that natural convection does not occur with horizontal freezers. As soon as the door to the upright freezer is opened, natural convection takes place immediately. The cold, heavy air falls out and is replaced by ambient air. These principles of natural convection are also used in home heating systems that have radiators or baseboard convectors. Heating of water in glass vessels such as coffee pots on stove tops, or beakers of water in the laboratory allows one to see the effects of natural convection. If there is a light behind the vessel refraction of the light by the varying densities enable the natural convection to be seen. The less dense fluid moves upward in a cell-like fashion. This phenomenon of refraction of light is used to experimentally detect natural convection and to even measure local densities. The employment of optical systems based on use of index of fraction has been termed "Schlieren Method".

Vertical fluid systems undergo natural convection as soon as a density gradient is established, but horizontal systems heated from below have a finite time before the onset of natural convection[2]. The layer of fluid becomes unstable only when the Rayleigh number exceeds the critical value of 1000. While we shall not be using it, some understanding of natural convection and the factors involved can come from the definition of the Rayleigh Number, Ra:

$$Ra = \frac{g\alpha|\beta|}{\kappa\upsilon}d^4$$

g = acceleration of gravity
d = depth of fluid
β = temperature gradient
α = coefficient of volume expansion
κ = thermal conductivity
υ = kinematic viscosity.

The shape of the cells moving upward are affected by the size of the vessel and the lateral boundaries[2], i.e. the shape of the vessel.

The effects of gravity relates to the density giving rise to buoyancy forces[3] Direct heating or cooling is only one of the ways to change local density. A second way is indirect heating caused by radiation absorption (including illumination in the case of colored solutions), where the radiation excites the absorber which then converts the excitation energy to thermal energy by collisions. A third way to generate density differences is by locally changing concentrations of solutions as can occur with electrochemical processes.

In a later section we shall discuss electrochemical impregnation of a number of battery active materials which rely on convective processes. For now we shall look at one mechanism used to explain the process and see the interaction of an electrochemical process at planar electrodes and how natural convection may become important. Streinz and coworkers[4] investigated the cathodic deposition of $Ni(OH)_2$ from an acidic $Ni(NO_3)_2$ solution. This sort of process had previously been put into production in a number of industrial plants using porous Ni plaque as the substrate in order to fabricate nickel oxide electrodes for secondary batteries. The process is important because, when properly done it results in improved weight, volume and life characteristics. The convection processes within the porous structures are quite different than those at planar electrodes, but we shall concentrate here on the need for consideration of natural convection.

The concentration range of the $Ni(NO_3)_2$ solution used by Streinz was 0.01 to 3M, while the industrial process run at about 2M. The electrodes were run vertically and were spaced apart. Deposition was on the planar electrode which was quartz coated with Au. The anode conductor is Pt. Inert anode conductors are favored so that the anodic process is evolution of oxygen gas. The interelectrode space is great enough so that stirring, forced convection, does not occur around the cathode because it was unwanted. The primary electrode processes may be represented as:

cathode: $\quad HNO_3 + 6H_2O + 8e^- \rightarrow NH_4^+ + 9OH^-$

anode: $\quad 4H_2O \rightarrow 2O_2 + H^+ + 8e^-$.

As this reaction proceeds there is a local pH change at each electrode. When the local pH at the solid cathode reaches a value about 9, $Ni(OH)_2$ begins to precipitate upon the metallic substrate as follows:

at cathode: $4.5Ni^{++} + 9OH^- \rightarrow 4.5Ni(OH)_2$,

so that the net cathode process is:

$HNO_3 + 6H_2O + 4.5Ni^{++} + OH^- + 8e^- \rightarrow NH_4OH + 4.5Ni(OH)_2$.

The consumption of m moles of Ni^{++} salt locally also consumes $\dfrac{4.5}{6} \times m$ moles of water. However, the starting amount of water in molarity was about 27 times greater than the Ni^{++} salt so that the local solution concentration is decreased. At vertical electrodes convection due to a buoyancy gradient is expected. When this occurs, there will be a further dilution of the pH change which affects the rate at which $Ni(OH)_2$ would deposit. For this investigation to avoid confounding by natural convection a change of orientation necessary. The experiment should be done horizontally; furthermore, the anode should be an inert screen so that there is no impediment to escape of the oxygen bubbles. Next, one must run the experiment under conditions where the Rayleigh number is less than the critical value of 1000, or alternatively, verify a quiescent electrochemical system by use of Schlieren methods.

1   S.Chandrasekhar, Phil. Mag. *43*, 501, (1952).

2   R. K. Soberman, J. Appl. Phys. *29*, 872 (1958).

3   Y. Jaluria in "Handbook of Single Phase Heat Transfer" Ed. S. Kahac et. al., Wiley-Interscience, New York (1987).

4   C. Streinz, J. Electrochem. Soc. *142*, 1084 (1995).

5   H.N. Seiger and V.J. Puglisi, Proc. of 26th Power Sources Symposium, Atlantic City, NJ, p. 115 (1974).

XI

Common Electrolyte Paths

Common electrolyte paths can come about through oversight in design, or careless handling of multicell electrochemical system, by inadvertent ejection of electrolyte during overcharge or overdischarge, or by design to maintain electrolyte concentration between the electrodes using flowing electrolyte. We shall deal with the case of design where there are manifolds for the inlets and outlets resulting in a common electrolyte path between every cell in the battery which otherwise are in series configuration.

Leakage currents are present in the common electrolyte paths and waste energy and decrease available capacity of the battery in a nonuniform way so it is important to understand, quantify and control them. Leakage currents are of importance in common pressure vessel configuration of metal oxide/hydrogen systems, remotely activated batteries, some redox and zinc/halogen batteries as well as in the chlor-alkali industry.

Most of the previous approaches to this problem entail calculations using electrical analogs, but there were also several experimental measurements reported[1,2,3]. One reported measurement of the leakage currents uses annular magnetic field detectors around the cell inlet or outlet channels which lead to the common manifolds:. The second kind of experimental measurement involves the placement of unpoised electrodes in the inlet or outlet channels and deducing the leakage currents from the voltage[1,4]. Poised electrodes have electroactive materials in both oxidized and reduced forms so that they resist voltage change when current is passed. Unpoised electrodes can exhibit voltage change with current which may be exemplified by Pt that evolves oxygen upon anodization or hydrogen upon cathodization.

Each of these methods is questioned with regard to yielding the information sought about leakage currents. The calculations all have used Kirchoff's current law which implies that the electrochemical systems are analogous to electrical systems. The value of analogy is limited by the physical and chemical differences between electrical and electrochemical systems. Four differences that restrict the use of the usual form of Kirchoff's law are : (i) electrochemical systems have two charge transfer interfaces with oxidation processes at the anode and reduction processes at the cathode, (ii) the rate at which the anodic process occurs is equal to the rate at the cathodic processes occur; the reactions are coupled, (iii) the natures of the charge carriers are different on the two sides of each electrode/electrolyte interface; ionic conduction in the electrolyte and electronic conduction in the solids; and (iv) the Gibbs energy change associated with the charge transfer processes constitutes the fundamental voltage source, but requires a pair of electrodes.

The experimental measurement with unpoised electrodes is questioned for several reasons of which the most obvious concerns the nature of the interaction of the "inert" metal with the electrolyte. While the measurements are being made, uncertainties are introduced by this interaction which prevent leakage currents from being deduced. The interactions are a result of drawing some current to make the measurement causing the unpoised electrodes to change potential. The interpretation of the resulting voltage is complicated.

Recognizing that, in general, even on open circuit, the cathode of an interior cell discharges with the anode of a contiguous cell while the anode discharges with the cathode of the other adjacent cell, the current flow through the inlet or outlet channels is the net difference of the discharge rates of the two sets of electrodes. Hence, the significance of channel current measurements[1,2,3] is uncertain and requires resolution.

A series of experiments were run which represent a different approach to measuring leakage currents. These measurements were made first with the silver oxide/zinc battery and then with the nickel oxide/cadmium battery systems. The nickel oxide/cadmium could, in the procedures that evolved, be measured during charge as well as on open circuit and discharge. The final experiment in this series was done to ascertain the interaction of unpoised electrodes with the electrolyte when making current measurements.

The rate of capacity decrease of the two sets of electrodes in a battery cell is equal to the current that passes through the intercell connector. This current may be composed of a component that passes through the external load and of another component involved with internal leakage currents. The anodes and cathodes within any one cell may have leakage currents that are different. The measurement of the currents through the intercell connectors is reported upon here.

a. Initial Investigation.

Figure XI-la is a representation of a bipolar battery having a common electrolyte. If one bipolar set is removed conceptually along with its electrolyte we have the representation shown in Figure XI-lb. In this representation there is the intercell connector along with the anode of one cell and the cathode from the contiguous cell. If next we conceptualize a slicing of the intercell connector so that the electrodes of the contiguous cells become two independent electrodes we

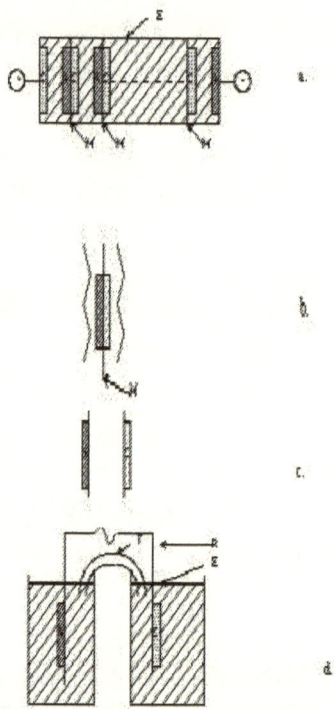

a. Schematic drawing of a bipolar battery having a common electrolyte.

b. Removal of a bipolar set having the anode A from one cell, the cathode C from a contiguous cell with the intercell connector M and the electrolyte E common to both electrodes.

c. Representing the splitting of the bipolar set of b into separate anode and

d. Recreating the bipolar set electronically with a wire R and ionically with an electrolyte bridge T.

Figure XI-1. Schematic showing the transition from a bipolar battery to the bipolar model.

arrive at the conceptualization shown in Figure XI-lc. Next, place these two electrodes into separate vessels and reintroduce the intercell connector with a wire R and place a tube, 'T', between the two vessels When the tube T is filled with electrolyte we have reintroduced the electrolyte path between the two electrodes reaching the arrangement shown in Figure XI-ld which is now a model of the bipolar set reached by conceptual slicing through the intercell connector. The wire R may be a shunt or an ammeter for leakage current measurement, and the geometry of tube T may be varied to simulate channel and manifold impedances. One way to vary the ionic impedances is to replicate the actual tube T several times over. The experiments[5] used "T" tubes made of 4 mm glass tubing of 11.5 cm lengths. The electrolyte was 5M aqueous KOH, the specific conductivity of which is 0.55 mho/cm obtained from the International Critical Tables. The calculated impedance of the electrolyte bridges from these data is 166 Ω. From 1 to 3 bridges were employed in parallel so that the range of intercell impedances was from 55 Ω to 166 Ω.

Part of the investigation was to measure distributed currents. The distributed currents are defined as those leakage currents between cells that are not adjacent. This was done by rerouting wiring so that the individual cells in the sample multicell battery were not physically moved while electrically they were separated.

One of the experimental methods for shunt current measurements found in the literature[1,4] was the use of unpoised electrodes. Unpoised electrodes interact with the electrolyte when attempts are made to measure voltage as shown in Figure XI-2. The flow of electrons through the voltmeter requires the current flows shown in Fig. XI-3. There are 4 charge transfer interfaces, hence are 2 sets of electrodes. Tracing the ionic and electronic compartment is an adsorbed hydrogen electrode contrasting with the one in the AgO compartment being an

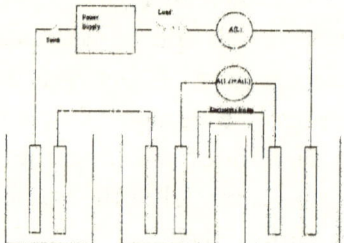

Figure XI-2. Three cell battery simulating a bipolar electrode, anode cell 1 with cathode cell 2 with common electrolyte
brought about by use of an electrolyte bridge.

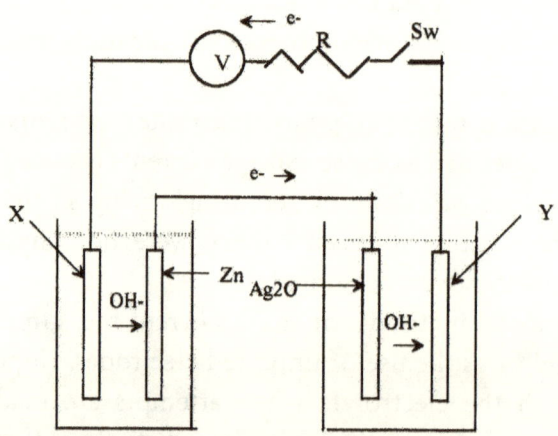

Figure 3. Effect of Battery Electrodes on Unpoised Electrodes.

XI - 7

adsorbed oxygen electrode. It is obvious that the metal electrode in the right hand chamber became slightly negative with respect to the AgO, and the other metal electrode becomes slightly positive with respect to the zinc electrode. The meter would then measure the voltage of the AgO/Zn cell plus the voltage across an adsorbed O_2/adsorbed H_2 cell. In the reported experiments the voltage of the battery cell was 1.68 volts and of the combination of Figure XI-3 with the unpoised electrodes averaged 1.82 volts for an excess voltage of 0.24 volts. With two cells included the excess voltage was 0.37 volts. Unpoised electrodes are, therefor, unsuitable for voltage drop measurements in electrolyte without compensation for the interactions that do occur.

The second experimental method to determine shunt currents found in the literature measured channel currents. The net current in a channel (as opposed to the manifold) responsible for commonalty of electrolyte is the difference in bipolar shunt currents involving the two sets of electrodes within the cell. The difference can be otherwise unrelated to the magnitude of the rate at which charge is lost by bipolar shunt currents.

The investigation using the experimental setups of Figures XI-2 and XI-3 emphasize the importance of the role played by contiguous electrodes. The rate of electrochemical capacity consumption to leakage as well as discharge into an external load is the same for the pair of electrodes electronically connected together, and interestingly, the set of electrodes within the same cell may be undergoing electrochemical discharge at different rates.

Figure XI-4 is useful for identifying some of the electrochemical circuits in systems having common electrolyte. When bridge 5/4 is the only bridge in place, electrodes κ_5 are the only ones that interact to yield a leakage current. In an alkaline electrolyte where the ionic flux is a hydroxyl ion flux, the origin of the flux is the cathode with flow to the anode. When bridge 4/3 is also installed the interaction of cathode κ_4 and anode is permitted. This flux flows opposite to that between κ_5 and it is in the channels that the interaction of κ_5 and α^3 becomes manifest. When all the currents between κ_5 and α_4 are equal to that between κ_2

and α_1 but these currents are less that between κ_4 and α_3 and κ_3 and α_2. The magnitude of the leakage rate is equal to the ionic transport rate through allowable paths which depends upon the driving force and the ionic impedance path. The driving force is the sum of the half cell voltages measured across the charge transfer interfaces for each electrode in the electrochemical circuit. The electrochemical circuit tracing includes the identification of an anode/electrolyte interface, a cathode/electrolyte interface, an electrolyte path between the participating electrodes and a load pathway. The ionic impedance path in some common electrolyte systems may be complicated. The effective impedance may be ascertained by modeling or by simulation with resistance paper or separator made conductive with electrolyte. If the leakage path between each cell in a battery is the same, the results of these findings can be expressed mathematically.

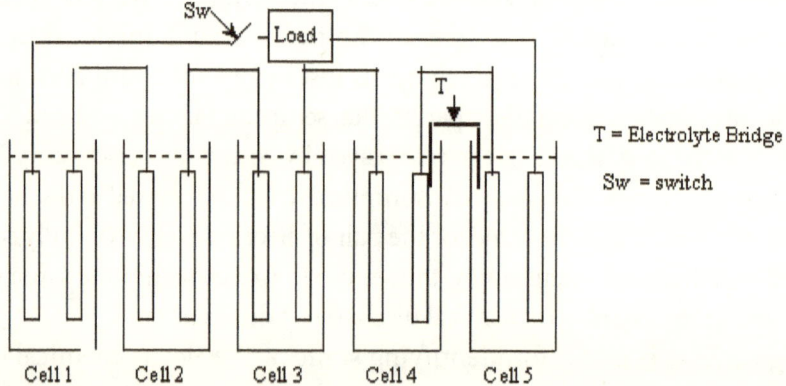

Figure XI-4. Diagram of a Five Cell Battery for Study of Effect of Common Electrolyte Pathways.

b. Mathematical Model

The mathematical model for leakage currents espoused here had not been previously published is the open literature. The approach taken now is that of a literature article rather than a pedantic review. What shall be done first is to set up the mechanism for the leakage currents, use an electrical analog followed by the mathematical expressions which are then solved. The scientific method requires a testing of the mechanism. To test the mechanism the derived expressions shall be applied to some other data in the open literature as well as data in a patent on the subject of protecting against leakage currents. Most importantly, a battery specially designed for this purpose was made, measured and results compared to those predicted by the model.

1. Theory

The experimental work establishes that the anode of one cell discharges at the same rate as the cathode of the contiguous cell. In a battery of N cells delivering power into a load, the power is IV_{bat} where I is the load current and V_{bat} is the voltage of the battery which is equal to the sum of the individual cells at the effective current passing through them, $\sum V_x$, so that polarization is taken into account.

Each cell consists of an anode subjected to a current $I + j_N$ where is the leakage current of a particular cell. The term cell, is used in several different ways which would lead to ambiguity and we must carefully identify them. A cell in an electrochemical system has an anode and a cathode. If the reactions are spontaneous, $\Delta G < 0$, it is a battery cell and the battery consists of N such cells in series. With common electrolyte paths, the anode of any one included cell is connected to the cathode of the contiguous cell by the intercell connector. Thus the Xth cell of a series string has an anode α_x and a cathode κ_x. The leakage cells are the anode a_x and the cathode of the contiguous cell κ_{x+1} also κ_x and α_{x-1}. The leakage current passing through the leakage cell a_x, k_{x+1} and the leakage cell α_{x-1}, κ_x need not be equal. The load current I is, however, the same for all cells. The rate at which electrochemical processes occur, except for symmetry, differs for all leakage cells and may be written as $I + j_{x,x+1}$. If the system is on open circuit, then the end electrodes (the terminal electrodes in a battery) have I=O, but all other electrodes are undergoing spontaneous leakage currents through the channels and the manifold segments, the rate of the process depending upon location. Figure XI-5 is a diagram of a 5 cell system on open circuit. Assume an aqueous battery, then each leakage cell behaves in way similar to that shown for the leakage cell we shall call α_2, κ_3.

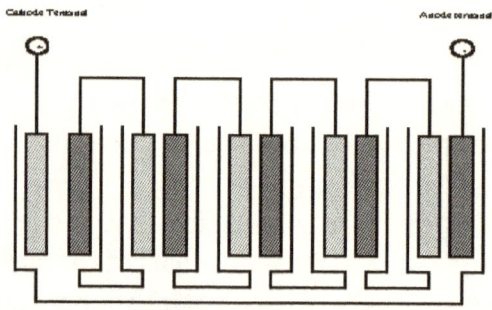

Figure XI-5. Diagram of a Five Cell Battery with Common Electrolyte

The leakage current is governed by the effective voltage of the polarized electrodes (α_2, κ_3) and the impedance of the intervening channels and the manifold segment. The polarized voltage is taken as $V^o_{cathode} - V^o_{anode} - S(I,t)$ where $S(I,t)$ is an internal resistance term that may or may not be constant depending upon the particular system. In Figure XI-3 the cathode material is represented as MO and the anode material as A. The electrochemical half cell reactions are given as:

cathode: $MO + H_2O + 2e^- \rightarrow M + 2OH^-$

anode : $A + 2OH^- \rightarrow A(OH)_2 + 2e^-$

The electrons are transported through the intercell connector and the OH⁻ ions through the electrolyte. Charge transfer takes place at the interfaces of each active material and the electrolyte.

The leakage current in the cell α_2, κ_3 may be written as:

$$\left(V_{k3}^{o} - V_{\alpha2}^{o}\right) - \left(I + j_{3,2}\right)S(I,t) - 2\left(j_{3,2}R_{C}\right) - j_{3,2}R_{M} = 0$$

$$j_{3,2} = \frac{\left(V_{\kappa3} - V_{\alpha2}\right) - IS}{2R_{C} + R_{M} + S}$$

or

where we made the current and time dependency of the internal resistance implicit in the S term. This expression is correct only when all the intercell connections are removed except for the one between electrode α_2 and κ_3,. If these other intercell connectors are still in place, then we must account for the interaction of adjacent cell leakage currents in the channels. Since the channel currents of contiguous cells are in opposite directions, the voltage drop in the channels is deceased. It is this channel current interference that brings about an interaction that results in greater battery cell voltage drops and greater leakage currents for centrally located cells, and the converse, leakage currents for the cells closer to the terminals are less.

Expressed in the form above, the analogy to electronic circuits is obvious, and we shall set this up mathematically so that the interaction between cells may be ascertained along with the consequent cell voltage distribution. Representation as a battery is shown in Figure XI-5. The polarization function S is depicted as next to the cell voltage when, in actuality, it is within the electrodes; part due to the anode and the remainder due to the cathode. The leakage cell voltage is always an electrode difference whether polarized or not, and we can solve for

$$V'' = \left(I + j_{x,x+1}\right)S = V_{\kappa x} - V_{\alpha(x+1)}$$ once S is known. In the case of a lead acid battery is reasonably constant, while in the cases of the silver oxide/zinc and nickel oxide/cadmium batteries the voltages are a function of both current density and the state of charge. Tables can always be created by making voltage difference measurements using poised electrodes that can either (1) be immersed in the electrolyte of cell x and read between the intercell connector attached to anode α^x and

the reference electrode. Then move the reference electrode (2) to the electrolyte of cell x+1 and measure from this reference electrode to the same intercell connector thereby obtaining the relative voltage of cathode κ_{x+1}.

The difference of these measurements yields $V_{x,x+1}$ the leakage current driving force. Alternatively, a pair of poised reference electrodes may be placed in the electrolytes of contiguous cells and measurement of the leakage cell voltage made, which is particularly good to determine polarization of these electrodes as a function of current density and state of charge.

The leakage current of the first leakage cell is written as:

$$(1) \quad \left(V_{\kappa 2} - V_{\alpha 1}\right) = j_{1,2} \times R_C + j_{1,2} \times R_M + \left(j_{1,2} - j_{2,3}\right) \times R_C$$

and included leakage cells:

$$(2) \quad \left(V_{\kappa x+1} - V_{\alpha x}\right) = \left(j_{x,x+1} - j_{x-1,x}\right) \times R_C + j_{x,x+1} \times R_M + \left(j_{x,x+1} - j_{x+1,x+2}\right) \times R_C$$

and for the last leakage cell:

$$(3) \quad (V_{\kappa N} - V_{\alpha N-1}) = \left(j_{N-,N1} - j_{N-2,N-1}\right) \times R_C + j_{N-1,N} \times R_M + j_{N+1,N} \times R_C$$

These equations can be simplified and rewritten as:

$$(1a) \quad \left(2R_C + R_M\right) \times j_{1,2} + \left(-R_C\right) \times j_{2,1} = \left(V_{\kappa 2} - V_{\alpha 1}\right)$$

$$(2a) \quad \left(2R_C + R_M\right) \times j_{x,x+1} + \left(-R_C\right) \times j_{x-1,x} + \left(-R_C\right) \times j_{x+1,x+2} = \left(V_{\kappa x+1} - V_{\alpha x}\right)$$

$$(3a) \quad \left(2R_C + R_M\right) \times j_{N-1,N} + \left(-R_C\right) \times j_{N-2,N-1} = \left(V_{\kappa N} - V_{\alpha N-1}\right).$$

Equations 1a through 3a along with the previously discussed relationship between the effective internal resistance may he solved to ascertain the intercell leakage currents and the effect upon the distribution of battery cell voltages in a series string.

Batteries are devices in which energy is stored and the so-called discharge is removal of energy from the battery. The time rate of removal of energy is the power delivered by the battery. The power is the current in any leg multiplied by the impedance or resistance in that leg. Power is dissipated in the external load as well as in internal impedances and both must be accounted for by applying the conservation laws.

The power delivered to the external load, I^b is given by $V_{bat}^2 / L = P_L$ where the delivered power is P_L and V_{bat} is the battery terminal voltage. However, V_{bat} may also be written as:

$$V_{bat} = \sum_{x=1}^{N} \left(V_{\kappa x} - V_{\alpha x} \right)$$

The load current I is either measured or calculated as:

$$\left[\sum \left(V_{\kappa(x+1)} - V_{\alpha x} \right) \right] / L = I$$

The internally dissipated power is taken as the product of the voltage drop and the total current through the electrode pair and must exclude both end electrodes. This may be written as:

$$P_{int} = \left(V^o - V_{\kappa 1 \alpha N} \right) I + \sum_{x=1}^{N-1} \left(V^o - V_{\alpha \kappa \kappa (x-1)} \right) \left(I + j_{x,(x-1)} \right)$$

The inside losses are to be calculated from a relationship that is the analog of the well-known terms. The ionic fluxes through the impeding solution are identified and the proper sums taken.

The power dissipated in the channels having an impedance R is:

$$P_C = \left(j_{1,2} \right)^2 R_C + R \sum_{x-1}^{N-1} \left(j_{(x-1),x} - j_{x,(x+1)} \right)^2 + j_{(N-1),N} R_C$$

and recognizing that $j_{1,2} = j_{N-1,N}$, then

$$P_C = R_C \left[2 \left(j_{1,2} \right)^2 + \sum_{x=2}^{N-1} \left(j_{x-1,x} - j_{x,x-1} \right)^2 \right].$$

The power dissipated in the manifold is handled similarly:

$$P_M = J_{1,2}^2 R_M + \left(j_{2,3}\right)^2 R_M + ... + j_{N-1,N}^2 R_M$$

and simplifying:

$$P_M = \left(\sum_{1}^{N-1} j_{x,x-1}^2\right)$$

The power dissipated within the battery which is responsible for the internal temperature rise, P_D is given by the sum of all the inside and internal terms:

$$P_D = [\left(V^o - V_{\kappa 1, \alpha N}\right)I + \sum_{r=1}^{N-1}\left(V^o - V_{\alpha x, \kappa(x+1)}\right)\left(I + j_{x,x+1}\right)] + R_M\left(\sum\left(j_{x,x-1}^2\right)\right).$$

At this point we have all the equations necessary to determine the distribution of leakage currents, the distribution of cell voltages and the power dissipation in both the external load and in the internal paths. The solution is given in the appendix.

2. Testing of the Theory

A six cell lead acid battery was built to test the mathematical model based in the new theory of common electrolytes. The plates were 4.5 cm wide and 2.8 cm high. Intercell connections were made using 1 ohm resistors. The aqueous H_2SO^4 electrolyte had a density of 1.223 g/cm^3 and the conductivity was taken as 0.838 mho-cm^{-1} from Langes' Handbook of Chemistry.

The channel dimensions from the region between the electrodes to the manifold corresponded to an area of 0.02 cm^2 with a length of 1.68 cm. The path impedance of the electrolyte in the channel was estimated as 99 ohms.

The distance between cells measured 1.04 cm. The manifold was drilled out of Noryl to have a cross sectional area of 0.126 cm^2 from which the manifold segment impedance was calculated to be 9.88 ohms.

By discharging the battery at several rates the internal impedance was found to be nearly constant at 2 ohms. Using an open circuit voltage of 2.15 volts per cell we have values from 5 of the 6 input parameters required by the BASIC program which are:

Input	*Value*
Number of cells in series	6
Channel Impedance	99
Manifold Impedance	9.88
Cell Open Circuit Voltage	2.15
Cell Polarization Slope	2

Measurements of leakage currents were made across the 1 ohm resistors behaving as shunts with the battery on open circuit. These intercell connections were made one at a time so that **N** was varied from 2 to 6. The experimental values are shown as points in Figure XI-6 while the calculated values are entered numerically near the corresponding points. It is obvious that the calculated values are in good agreement with the experimental values.

The model may now be applied to data previously given in the literature. The first selected is in a patent[5] by Zahn and his coworkers. They had made an electrolysis system and built in long channels as well as a manifold. They used a device to measure the magnetic field generated by the current in the channels (sic) and reported these values[6] for several electrolysis currents. The comparison of their experimental data with our calculations is not on leakage currents, but on channel currents since they are the data reported. We have to note that because of symmetry the channel currents decrease toward zero in the center of a series of cells, and increase again toward the other end of the stack. The magnetic field detector was not inverted so that the polarity reported by Zahn et. al. is negative. Now, understanding the process better our data are reported as positive values so that the comparison should be made on the basis of absolute values. The two sets of comparisons are shown in Table XI- 1 where reasonably good agreement is found.

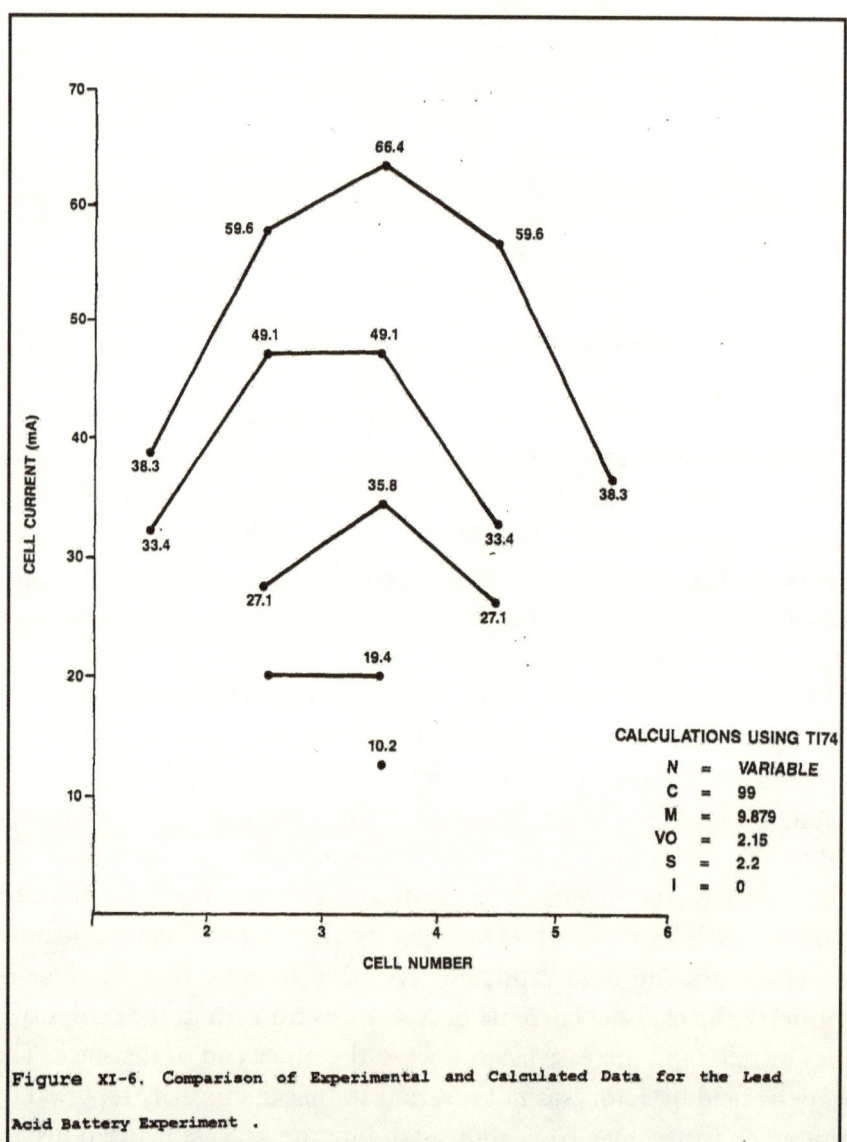

Figure XI-6. Comparison of Experimental and Calculated Data for the Lead Acid Battery Experiment .

The second example of literature data selected is a paper by White[7] and coworkers. This is interesting because they use the concept of a divided cell. Thus they have two input channels per cell, one on each side of the divider, and two manifolds. To effect of a comparison between their data treatment and that of the model just developed the data have to be put on the same basis, as was done with Zahn's data. In this case the input and output impedances are combined as a parallel set and the leakage currents summed. These data are given below as Table XI-2.

Table XI-1. Electrolyzer Experiments, Channel currents compared with Mathematical Model.

	Example 1, 124 mA 24 V	Calc'd.*	Example 5 420 mA, 25.4V	Calc'd.*
S^1	11	11.4	10	8.4
S_2	5	5.2	4.3	3.9
$S3$	2	2.3	1.8	1.7
S^4		1	0.85	0.7
S_5	0.31	0.3	-0.15	0.2
S_6	-2.4	0.3	-0.4	0.2
S_7		1	-1.35	0.7
S_8	-3.4	2.3	-1.4	1.7
S^9	-6.3	5.2	-4	3.9
S_{10}	-11.5	11.4	-10	8.4

*Program using data from Zahn patent including derived information: N=10, C=i28, M=80, VO=1.9, S=1.5, I=0.124 and 0.420, s 1M KOH =0.25 mho/cm and di-mensions given in Column 10 lines 49 ff of patent. Note that S_6 to S_{10} are tabulated as positive values whereas a measurement to the surrounding field magnetically would have a negative value.

c. Conclusions

The mathematical model based on a leakage cell consisting of an anode and a cathode in contiguous cells joined two-foldedly by an ionic common electrolyte path and electronically by an inter-cell connector was able to predict the leakage currents, cell voltages and battery voltage of a specially designed lead acid battery. Going beyond the prediction, two other sets of data from the literature was in concordance with the model. One was an electrolyzer system built and measured by Zahn and coworkers at the Exxon Research Laboratories. The other was an electrochemical device built and tested by White and his coworkers using the concept of divided cells resulting in parallel channels and manifolds.

Input Parameters

White et. al.[1]	Mathematical Model
$I_{T} = 0.1\,A$	$I_{T} = 0.1\,A$
$N = 11$	$N = 11$
$V_{o} = 1.0$ volts	$V_{o} = 1.0$ volts
$R_{c} = 3$ ohms	$S = 3$ ohms
$R_{t,\,in} = 1200$ ohms	
	$C = 545$ ohms
$R_{t,out} = 1000$ ohms	
	$M = 2.4$ ohms
$R_{t,out} = 6$ ohms	
I_e , mA	$j_{x,x+1}$, mA
5.79	5.8
10.35	10.4
13.74	13.8
15.98	16
17.09	17.1
17.09	17.1
15.98	16
13.74	13.8

10.35	10.4
5.79	5.8
0	-

The model can be used to determine the design of electrochemical systems which require a common electrolyte. This includes the effect of the number of cells in the unit, the shape and length of the channels and the manifolding, the effect of factors that impact effective internal impedance, the nature of the electrodes and , finally, the current required y the load.

Since there is a capacity decrease even on open circuit, one can conceivably compensate by having a graded capacity based on distribution of leakage current and elapsed time.

1   I. Rousar, J. Electrochem. Soc.,*116*,676 (1969).

2   M. Zahn, ety. al. U.S. Patent 4,197,169 (1980).

3   M. Katz, J. Electrochem. Soc., *125*, 515 (1978).

4   I. Rousar and V. Ceznar, ibid., *121*, 648 (1974).

5   H.N. Seiger, J. Electrochem. Soc. *133*, 2002, (1986).

6   M. Zahn, et. al., U.S. Patent 4,197,169.

7   R.E. White, et. al. J. Electrochem Soc. *133*, 486, (1986).

XII

Exchange Current Densities

There are isotopic exchange reactions involving different valence states which may be exemplified with the thallous-thallic couple:

$$Tl^{*+} + Tl^{+3} \Leftrightarrow Tl^{+} + Tl^{*+3}$$

where the radioactive isotope Tl^{204} is indicated by Tl^{*} and the stable isotope Tl^{203} is also present. This couple had been investigated by Prestwood and Wahl[1,2] and found to be slow under homogeneous conditions. When Pt was inserted into the solution the heterogeneous exchange reach equilibrium within the time required for separation of the two valence states. So that at 25°C the half time of 125 hours for the exchange reaction under homogeneous conditions increased so that isotopic equilibrium was reached within three minutes.

The reactions may be written as anodic and cathodic half-cells:

anode: $$Tl*^{+} \Leftrightarrow Tl*^{+3} + 2e^{-}$$

cathode: $$Tl^{+3} + 2e^{-} \Leftrightarrow Tl^{+}$$

If we concentrate on the heterogeneous reaction the resemblance to an electrochemical half-cell is obvious:

$$Pt \, / \, Tl^{+}, Tl^{+3}$$

Consider the conductive metal to be a reservoir of electrons. When a Tl^{+} ion collides with the conductor the other 2 valence electrons can be part of an adsorbed complex of $Pt - Tl^{+}$. Elsewhere on the surface of the solid Pt, a similar adsorption of Tl^{+3} can exist. With a dynamic process it is equally likely that the labeled $Tl*$ can separate either taking a pair of electrons or leaving behind a pair of electrons. Similarly, a Tl^{+3} ion that collides can split apart later as the original Tl^{+3} or capture a pair of electrons to leave the surface as a Tl^+ ion. The specific activity of the two valence states would reach equality (equilibrium) rather rapidly. Thus, some local regions of the $Pt / Tl^+, Tl^{+3}$ interface would be a localized anode and some other region would be a localized cathode. The rate of electron transfer could be expressed as a current. More particularly, since the rate of the electron transfer occurs without an applied potential, the electron transfer process could be considered to be an exchange current density[3.]

The isotopic exchange reactions are not chemical reactions in the ordinary sense that they are accompanied by a change of Gibbs energy. Such exchange processes are always occurring but are observable only by use of isotopic labeling. Thus isotopic exchange reactions and their rates which we considered to be exchange currents are characterized by the thermodynamic characterization $\Delta G = 0$ that for the process.

There is, in the field of electrochemistry, another "exchange current" which has to be defined and explained in the same way we have done with the isotopic exchange current. It started with the experimental Tafel equation of 1905[1,] but Gurney[2] did a theoretical derivation to obtain the same form. This equation is similar to the one derived in Chapter VI;

$$i = i_o \exp\left(\frac{-\alpha z F \eta}{RT}\right),$$

i is the current through the external circuit, i_o is the exchange current thought to occur when the overpotential η is zero. The term α is called the transfer coefficient while F is the faraday, R and T the gas constant and absolute temperature. The z term represents the number of electrons involved with the faradaic process. The value of i_o is determined by plotting logversus overvoltage h and extrapolating. In this case:

$$\log i = \log i_o - \kappa\eta$$

but when η=0 then,

$$\log i \big|_{\eta=0} = \log i_o$$

Immediately, when $i \neq 0$ there is current in the external circuit so that the Gibbs energy term is not zero. We are dealing with some non-spontaneous process such as electrolysis of water or metal plating. The argument of absolute single potentials is used again showing lack of definition for a system having only one "electrode". An isotopic exchange may take place when a metal is placed in a solution containing ions, but the two electrode processes do not occur in isolation.

It was shown earlier that for reactions which are not spontaneous the two terms of the equations can not be neglected, and when $\eta = 0$, no current flows through the external circuit and there is no net ionic current between the two electrodes.

1   Prestwood and Wahl, J. Amer. Chem. Soc., 70, 880 (1948).
2   Idem., loc. cit. 71, 3137 (1949)
3   J. Tafel, Z. phys. Chem., 50, 641 (1905).
4   R.W. Gurney, Proc. Roy. Soc, 134A, 137 (1931).
5   U. Falk and A. Salkind, Alkaline Storage Batteries, John Wiley & Sons, New York, 1969.

XIII

Lithium Underpotential Deposition

Inside some bipolar cells of battery stacks having lithium alloy anodes and molten salt electrolyte sometimes lithium nodules are found on parts of the metal collecting plates. These nodular deposits were suspected of causing short circuits within the cells. Because of this it became important to identify the mechanism for this growth so that it may be controlled or avoided. A number of mechanisms were conceived which did not stand up under scrutiny until an electrochemical process was proposed. However, such a mechanism must rely on an underpotetial mechanism since the $LiAl$ and $LiSi$ alloys have potentials about 300 mV less negative than metallic lithium. Similar deposition of Li in primary silver vanadium oxide cells were studied by Takeuchi and Thiebolt[1] Their experimental methods led to the concept of underpotential deposition since they reported that the integrated current and their titrimetry results were in substantial agreement.

It was striking that the integrated current, which flowed between the anode terminal and the stainless steel case, was related to the quantity of Li deposited. The experimental methods and the findings were similar to another underpotential process, the adsorption of hydrogen on elemental nickel surfaces[2]. It was decided to adapt these methods used in the investigation of hydrogen on nickel to the molten salt system with a lithium alloy anode. Since the electrolyte was needed in the experiment procedure a separator consisting of MgO particles and electrolyte crystals were included. A Mo collector was placed on the opposite side of the separator, and this assembly was put into a press modified for use in a glove box. The ammeter would be connected between the anode collector and the "cathode" collector after the system was heated. It would be interesting to compare the effects due to this electrical connection so that a band of metal was cut from the "cathode" connector, and to keep these apart so that one section is electrically floating while the other section is connected to the anode (through the ammeter) an insulator was needed. It was into the insulator that the elemental lithium migrated, as will be described fully later.

The apparatus consisted of a modified press with heated platens. The press was installed in a glove box having a dry controlled argon atmosphere. The press was necessary to apply a 35 kg force to the cell during testing. The Macor insulators were placed around the heated platens. The platens were covered with a molybdenum sheet so that the copper from which the heated platens were fabricated would not be in contact with the electrodes. A disk of Mo with a tab was laid onto the lower Mo sheet covering the lower platen and the anode was placed directly onto this collector. The anode and separator pellets were made by compression in a die using 13 tonnes of force for the individual anode and separator pellets. They were cold fused in the same die by applying a 38 tonne force. A second Mo collector with a tab was placed on the separator, this upper collector was sectioned as shown in Figure XIII-1. The sectioned Mo was isolated from the Mo covering in the upper platen by the insulator under test.

The anode was a mixed $LiAl, LiSi$ alloy and its potential was governed by the α, β phase transition of the LiA1 alloy. Both anode and separator pellets contained the electrolyte which was composed of $LiF, LiBr$ and $LiCl$ with a melting point of 425°C.

This assembly was similar to that used for Li alloy/metal sulfide batteries described by Papadakis[3,] see Figure XIII-2. The test assembly was brought to 480°C and maintained at that temperature for the duration of the test. The test cell was equilibrated for about 50 hours before an electrical connection was made. A two ohm precision resistor, which served as a shunt, was connected between the two Mo tabs and readings taken using a Keithley Model 175 DMM multimeter so that a record of current versus time was made. These manual readings were backed up by recording the same data on a Hewlett-Packard 3497A DAS and 9817 for program control. Data were taken for about fifty-five hours, after which the system was cooled and the components removed.

The components were individually place into water, heated and stirred and then titrated using bromcresol green. Sometime later the indicator was changed to phenolphthalein. Some of the later samples were placed into a desiccator along with a beaker of water and sealed. The desiccator was attached to an inverted cylinder so that gases released, when the desiccator was tipped allowing the water to attack the insulator, could be captured and the gas volume thusly determined. The aqueous solution remaining in the desiccator was also titrated. In this way we had three sets of data that could be compared: (1) coulometric data from the current-time integration, (2) titrametric data from the alkalization of water by the components tested, and (3) eudiometric data from the gas evolved.

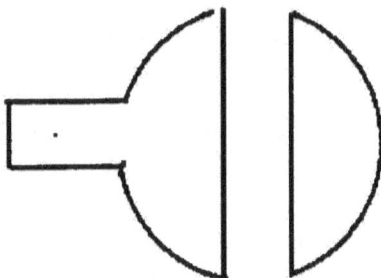

Figure XIII-1. Sectioned Molybdenum Foil Terminal.

The results of the first experiment in which the insulator was BN were at first perplexing, then became informative an intercalation of Li into the ceramic was deduced. Coulometry indicated 0.98 meq of Li, but the tabbed section of the Mo collector titrated to 0.07 meq and the control section titrated 0.08 meq. The missing material was subsequently sought in the ceramic. Visual examination revealed that the BN was swollen in the region of the split in the conductor, it was also discolored (blackened) and one edge started to crack. The backside showed no evidence of changes except for the edge crack. Titration of the ceramic indicated that the missing Li was in this material and that some time is necessary for it to leach out, see Figure XIII-3. The total amount of basicity by titrimetry exceeded the value obtained by coulometry. Subsequent testing of freshly machined BN showed that when this material is placed into the hot water there is hydrolysis of BN to H_3BO_3 and NH_3. After the hot water treatment the surface of the ceramic flaked off and felt like the slippery particles of H_3BO_3. The difference in results between titrimetry and coulometry is due to the basicity caused by NH_3. When the data are corrected for hydrolysis on the basis of surface area the agreement between coulometry and titrimetry was good.

While the form of the lithium present in the BN is not known, it was recognized that this work constitutes a procedure for testing high temperature insulators for the molten salt batteries. The presence of Li in the BN and the swelling of BN leads to the suspicion that the metallic Li is intercalated in the BN. DC resistance indicates no conductivity of the insulating material so that it is still performing its prime function in the battery. There had been a fear that if the material became electronically conducting a short circuit would arise. The presence of an elemental metal such as lithium would cause the onset of electronic conduction at a high enough concentration. Having it as an ion in the host would be no problem since ionic conductivity is already present with the molten electrolyte and discharge could not take place with a set of charge transfer processes.

Mica had been used in some cells as a high temperature insulator. A piece that had been incorporated into a battery was supplied by colleagues for testing in our apparatus. The coulometric results indicate 7.3 meq of *Li* deposition while

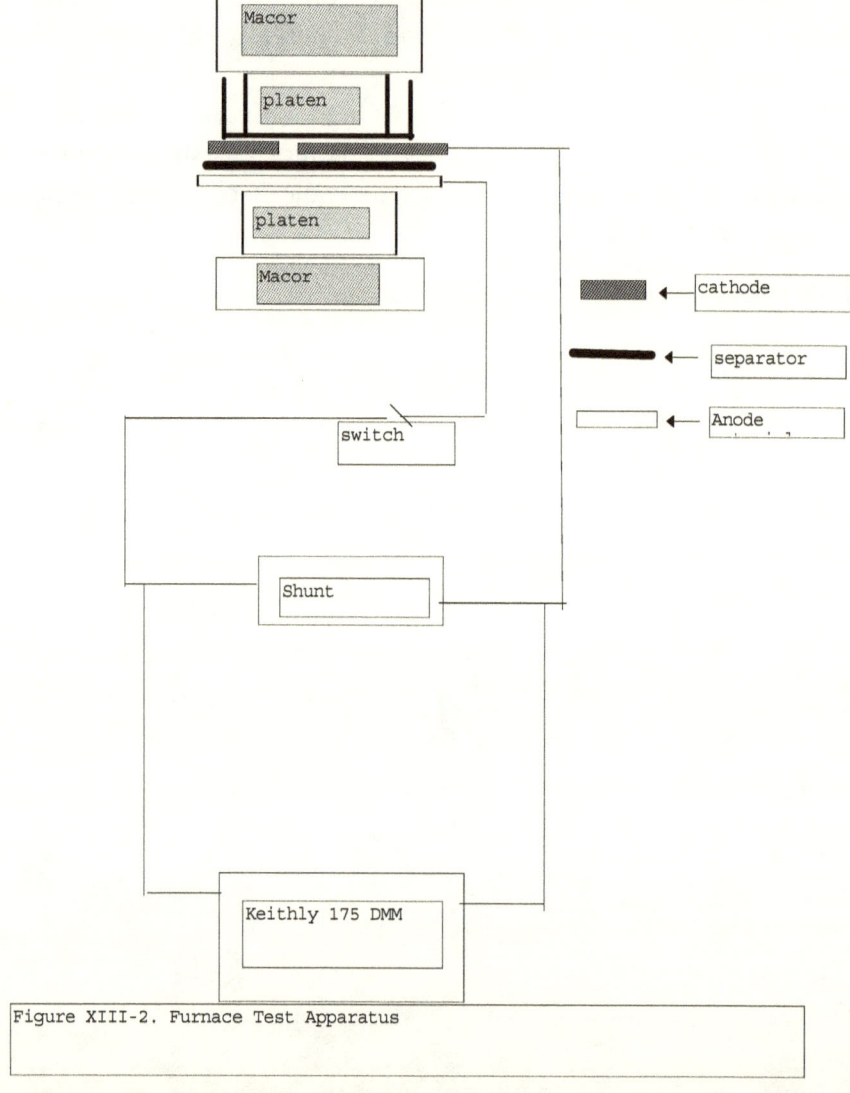

Figure XIII-2. Furnace Test Apparatus

the titration gave a much larger value, 22.1 meq. A new sample of mica was taken and run, but this time the gas was collected from the reaction between the mica and water in the desiccator. In this case coulometry and titration were in good agreement, but there was a problem with the gas collection. When the water and the mica were allowed to react, the reaction was so rapid that there was ignition with a blue flame. The lithium was obviously metallic and reacted as if it was finely distributed. The ignition, of course, resulted in an oxidation of the released hydrogen with the atmospheric oxygen and the gas volume results were low.

Samples of the materials as well as the separators and the electrolyte were used to run blanks. These materials, with the exception of the BN did not cause the phenolphthalein indicator to turn pink. They either were insoluble or hydrolyzed acidic. The peak current was over 7000 μA . When this material was placed in water it fizzed and disintegrated yielding floating black particles. It should also be noted that the AD998 broke into several pieces, showed considerable darkening and left some matter stuck to the Mo . It is possible that the Li deposited in the AD998 all became attached.

to the Mo as a buildup. The backside of the sample remained white. The darkening of the solution interfered with the titration. The behavior of this pure material was much different from that of the less pure AD94 MgO obtained from the same source. There is no reason ad initio to expect only one mechanism to be involved when the circuit is closed on the apparatus. Several processes were conceived which are:

> Underpotential deposition
> Ion Exchange
> Shuttle mechanisms
> Diffusion
> Intercalation.

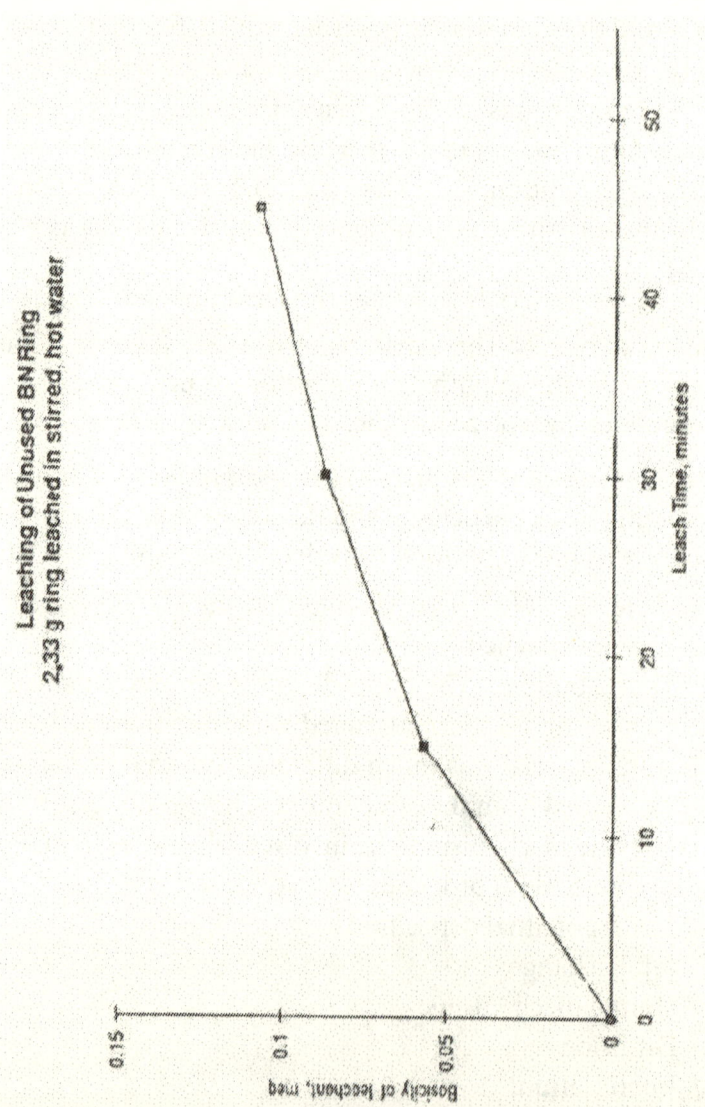

Figure XIII-3. Leaching of Unused BN ring. 2.33 gram final leached in stirred,

These processes are considered as follows.

a Underpotential deposition

The anode interface consists of molten electrolyte and the solid Li alloys. The potential is governed, in part, by the oxidation of Li to Li^+ ion as is known from other experiments. The potentials for the oxidizable materials which are identified are:

$$Li \rightarrow Li^+ + e^- \qquad -3.05 volts$$
$$Al \rightarrow Al^{+3} + 3e^- \qquad -1.66 volts$$
$$Si + 4F^- \rightarrow SiF_4 + 4e^- \qquad -1.2 volts$$
$$3I^- \rightarrow I_3^- + 2e^- \qquad +0.54 volts$$

The Li / Li^+ potential is sufficiently negative to maintain the other species in the reduced form. These considerations should be sufficient to establish that the anodic process is Li oxidation.

The coupled cathode process is more difficult to identify since the "cell" was not built with a readily reducible material. The molybdenum is already reduced[4]. The halides are also reduced. One of the materials tested was a cast, fired LiF board. It is only the ion of this material and of the electrolyte that is reducible. In which case we may write:

$Li^+ + e^- \rightarrow Li$ and this is, therefore, an underpotential deposition and the half-cell may be written as:

$$Li^+ X^- \| Li, Mo$$

The Li deposited upon Mo may diffuse into the structure of the LiF. As part of the experimental work it was shown (1) that the titer for LiX electrolyte is zero, and that (2) the control board of treated, cast LiF also had a zero titer. If these reactions and results are combined to characterize a complete electrochemical system the following is obtained:

$$Mo|Li(Al, Si)\|Li^+ X^-\|Li|Mo$$

a system analogous to other underpotential processes such as hydrogen adsorption in supercapacitors[5,] lead on gold[6] and the adsorbed hydrogen electrode[7.]

b. Ion Exchange

Mica being chemically similar to vermiculite may undergo ion exchange reactions as reported by Maraqah and his coworkers[8.] A heavier metallic ion in the mica structure, such as K^+ and Mg^{++} ions may undergo an ion exchange reaction with the molten electrolyte. When placed in water for leaching, the protons of water would then exchange with Li^+ *the* ions, i.e.:

$$Li^+ + K^+ - mica \rightarrow K^+ + Li^+ - mica$$

$$Li^+ - mica + H_2O \rightarrow LiOH + H^+ - mica \quad \text{LiOH + H}^+ \text{-mica}$$

The leachant becomes alkaline and is titrable. In this way an alkalinity greater than that equivalent measured by coulometry may be encountered. One of the overcharge shuttle mechanisms promulgated at ANL[9] for Li bearing batteries is rewritten as follows:

negative interface: $Li^+ + e^- \rightarrow Li$ (charge)

$$Li^+ + e^- \rightarrow Li \text{ (dissolution)}$$

positive interface: $Li_2^+ \rightarrow 2Li^+ + e^-$ (discharge)

When these are added together, there is no net reaction which is indicative of an effective shuttle mechanism. In the event of a shuttle mechanism operating we would have to find a lesser quantity of Li metal at the "cathode" region than indicated by coulometry.

c. Diffusion

A lithium diffusion process may not be "stand-alone", but accompany other processes. Thus lithium as the element dissolves in the electrolyte from the anode and diffuses away from the source. Nature tends to make the solution uniform in concentration[10.] When the dissolved, elemental LI reaches the "cathode", nothing further is expected to happen since there is nothing reducible.

If Li can enter the structure of the insulator, then a diffusional gradient is set up and deposition at the insulator continues. In such a case the quantity of Li found by analytical techniques can exceed that measured coulometrically. In this instance, as differing from ion exchange, the Li is metallic and behaves as does the metal when placed in water. If the metal is transported from the anode via a dissolution-diffusion process then titrimetry yields vales greater than coulometry. If the transport is via UPD which is then followed by diffusion into the insulator, titrimetry and coulometry should yield the same values. 5. Metathetic reactions

With a simple metathetic reaction there is a similarity to the an ion exchange reaction, except that a valence change is involved. Using alumina as an example, the metathetic process is:

$$6Li + Al_2O_3 \rightarrow 3Li_2O + 2Al$$

This sort of process would result in two effects. The first is that upon subsequent treatment with hot water $LiOH$ would form:

$Li_2O + H_2O \rightarrow 2LiOH$, and this would cause the titrimetric determination to exceed the coulometric measurement. The second result could be a discoloration of the alumina due either to the presence of Al particles or due to the introduction of defects into the structures of the $Li - Al$ oxides.

MgO was not tested, but there is a possibility of an analogous metathetic reaction between elemental Li and magnesia.

d. Intercalation

Intercalated chemicals are those where some specie is integrated into the structure of the host compound. It most usually is accompanied by an oxidation-reduction process, however, that would not be the case in the capture of Li, but an important characteristic of the intercalation process is a volume change of the insulator.

If intercalation is involved, the "cathode" molybdenum may not have any significant quantity of by titrimetry but the insulator is swollen and may even have undergone a color change.

While there were 6 possible reactions reviewed, there was a seventh type of happening which could arise. When separating the 3 components from the "cell" which has anodic material, there was an attempt to prevent cross contamination by avoiding any material easily removed from the molybdenum pieces or the insulator. Hence, errors may be introduced whereby less Li was titrated than measured coulometrically. In one particular instance it is reasonably certain that a very small lithium nodule was removed.

Seven materials were investigated. To analyze the data a plot was made of the difference between titrimetry and coulometry measurements versus the coulometric measurement, and it is now shown as Figure XIII-4. A horizontal line was drawn through the zero intercept. Values which lie near zero are consistent with an Underpotential Deposition (UPD) mechanism. The graph shows seven values which differ significantly from zero. The major negative deviation corresponds to the used mica samples and the second is a large positive deviation corresponding to the Coors AD94 alumina. In both instances the coulometric charge is quite high. While from a fundamental viewpoint the identification of the mechanism responsible for the errors may be a source of satisfaction, such information would not be useful in designing a better battery component.

It is noted that in the case of "not previously used mica", that the gas formed when the insulator was placed in water was captured and measured. There was ignition demonstrating elemental Li in the insulator, but this ignition caused an erroneous low reading for the gas. The gas actually measured corresponded to 5.5 meq versus an expected 7.1 meq based on coulometry. It is deduced that about 1/4 of the hydrogen produced burned to water.

Boron nitride was another uniquely behaved material. In contact with water it hydrolyzed to H_3BO_3 and NH_4OH so that titration measures both the Li that caused a swelling of BN and NH_4OH the from the hydrolysis.

In an evaluation of ceramic insulators for use with lithium electrowinning cells Lauf and DeVan[11] investigated the effect of molten Li on a large number of ceramics and in four instances this investigation happened to have used the same or similar materials. Both sets of investigators found the Coors alumina, AD998,

TABLE XIII-1		Summary of Insulator Effect on Lithium Transport						
ID	Insulating Material	Coulometry, meq	Insulator Titre, meq	UPD" Mo, meq	Control Mo meq	Gas Collection, meq	Lowest Current mA	Deviation Coul. - Titre
LCB M2075	BN	0.98	1.23	0.07	0.08		180	-0.4
lcb m2076	mica,used	7.31	22.1	0.03	0.09		1.25	-14.91
LCB M2081	mica	7.1	7.05	0.05	0.15	5.45	2.75	-0.15
910 83-9 #10	LiF	0.21	0.71	0.09	0.05		135	-0.64
910 83-9 #8	AlN_3	0.94	0.2	0.08	0.04		220	0.62
910 83-9 #11	Alumina AD94	15.93	2.88	0.07	0.1		6200	12.88
910 83-9 #12	Alumina AD998	7.13	0	5.9	0.03		450	1.2

to have blackened. In the chemical test Lauf and DeVan were not able to recover the AD998 while with the electrochemical procedure it was recoverable. The electrochemical test upon Coors AD94 showed attack while the molten lithium chemical attack was so severe that the specimen was not recovered. The *AlN* samples survived the molten *Li* attack and the electrochemical method found only small amounts of the UPD entered the material. The chemical attack upon glass bonded mica yielded a purple smoke when heated, and in the electrochemical procedure the resin bonded mica reacted violently with water as would *Li* powders or so severe that the specimen was not recovered. The *AlN* samples survived the molten *Li* attack and the electrochemical method found only small amounts of the

UPD *Li* entered the material. The chemical attack upon glass bonded mica yielded a purple smoke when heated, and in the electro-chemical procedure the resin bonded mica reacted violently with water as would *Li* powders or alloys. In general although Lauf and DeVan used a chemical method the test results are in good agreement with the electrochemical procedure.

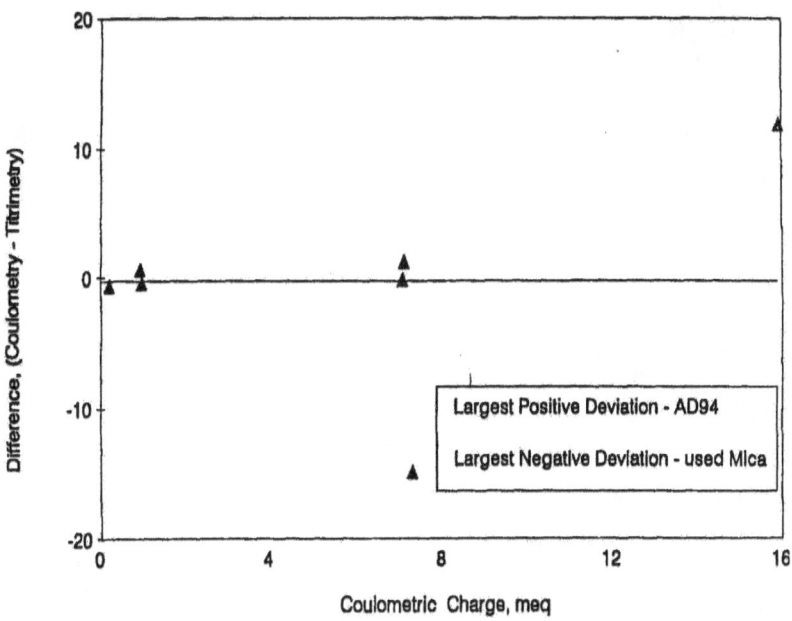

Figure XIII-4. Comparison of Charge and Titre.

1   E.S. Tekeuchi and W.C. Thiebolt, III, J. Electrochem. Soc., *138*, L61 (1991).

2   H.N. Seiger, paper delivered at the Electrochem Soc. Meeting, Fall 1963.

3   N. Papadakis, Development of a Bipolar Lithium/Cobalt disulfide Battery for Pulse Power, Proc. 34th Intern. Power Sources Symp. p.339, IEEE, New Jersey, 1990.

4   $Mo = Mo+3 + 3e- -.02$ volts

5   B.E. Conway, J. Electrochem. Soc. *138*, 1539 (1991)

6   F.G. Will and C.A. Knorr, Z. Electrochem., *64,258(1960)*

7   H.N. Seiger, et.al.,Proc. Ann. Power Sources Conf.,p.18,1964

8   H. Maraqah et.al., J. Electrochem. Soc. *138,L61(1991)*

9   L.Redey, Abstr.48, Electrochem. Soc. Mtg. Fall 1988

10 Uniformity is macroscopic and not microscopic as determined by study of radiocolloids.

11 R.J. Lauf and J.H. DeVan, J. Electrochem. Soc., 139, 2087(1992).

XIV

Electrochemical Impregnation with Nickel Compounds

The sintered plate nickel oxide electrode is capable of much higher discharge rates than the tubular, pocket plate or pasted plate varieties. The tubular and pocket plates supply an exoskeleton to the electrodes. The pasted plate lacks any skeleton, but the sintered plate provides a strong skeleton around the active materials. The nickel oxides involved in the varying valence states, corresponding to stored energy, differ in density and the volume change during the charge and discharge processes causes a hoop stress that leads to thickening of the structure. In turn, this thickening leads to destruction of the electrode, an electrolyte management problem and loss of capacity. In some cases there can even be a short circuit. The sintered plate confines the active material enhancing lifetime, and because of the distribution of the current collectors enables the electrode to be charged and discharged at very high rates.

The original sintering work was done in Germany during World War II at the I.G. Farben laboratories. This work was discovered by The U.S. Army at the end of the war and disclosed. A battery company was formed in 5he United States called the Nickel Cadmium Corporation of America and they hired Dr. Arthur Fleischer as Research Director with the task of bringing the nickel cadmium battery to market in the United States. Fleischer was responsible for the seminal work in the United States[1] The impregnation was carried out by immersion of the sintered plaque in aqueous solutions of the nitrate salts of nickel.

The aqueous immersion method used solutions that were of the order of 2 molar. If a vacuum is used to completely penetrate the pores with solution, then each cm^3 of voids contains only 2 millimoles of the Ni ions. The next step in the process is drying and then immersion in an alkaline solution to convert the nitrates to $Ni(OH)_2$. This step is accompanied by cathodization to hasten the conversion within the pores. The residual nitrates are electrochemically reduced and one readily smells the odor of the ammonia produced:

$$NO_3^- + 6H_2O + 8e^- \rightarrow NH_3 + 9OH^-$$.

Since the nitrate reduction is within the pores, the hydroxyl ions are produced where they are needed to precipitate the hydroxide:

$$Ni^{++} + 2OH^- \rightarrow Ni(OH)_2$$

deposition results in 2 millimoles of metal hydroxide and the plaque is, originally, about 80% porous, the capacity per unit volume of electrode is about 0.054 Ah/cm^3 based upon the anticipated 1 faraday per mole (F/M) electrochemical reaction. Commercial electrodes require capacities in the neighborhood of 0.4 Ah/cm^3; the impregnation process must be repeated several times. However, with the deposition and conversion of $Ni(OH)_2$ the available volume for penetration by the impregnation solution decreases. The Fleischer paper does indicate this effect of ever decreasing weight gain per impregnation cycle.

The immersion impregnation cycle is a labor intensive process and in the industry one finds anywhere from 4 to 12 impregnation cycles being used. There are several approaches to minimize the repetitive impregnation processes which are:

- Make a pasted or pressed electrode with the sinter.
- Increase the concentration of the metal nitrate solutions. The ultimate is to dissolve the salt in its own water of hydration. The hydrated salt melts at about 60°C.
- Chemically or electrochemically etch the nickel sinter.
- Cathodically deposit the hydroxide within the porous structure.

In the first case there is no exoskeleton structure and the volume changes occurring with the oxidation-reduction processes result in lessened cycle life. The second approach, molten salts have unique problems which require a separate treatment. The etching approach result in lessened capacity and a weakening of the structure of the plaque. The cathodic deposition has resulted in a single step (with respect to labor) impregnation that has a high capacity per unit weight and volume and also achieves an improved cycle life. The remainder of this chapter deals with electrochemical impregnation of oxygenated nickel compounds into the porous nickel structure.

One of the early workers in cathodic deposition of nickel hydroxide in a sintered nickel structure was Kandler[2] who used nickel nitrate solutions. He recognized the production of ammonia and the pH change responsible for precipitation. Hausler[3] not only duplicated Kandlers results but determined the rate of ammonia production in the mother liquor. When his results were corrected for the volume of solution trapped within the electrode - drag out - the ammonia production rate was in very good agreement with the theoretical production rate. However, these early works were done at room temperature and the impregnation levels were about half that of the commercially available electrodes.

McHenry[4] investigated the process at a variety of current densities and temperatures. McHenry's work is also considered seminal since it enabled others to subsequently improve upon the process. Among these were MacArthur[5,] Beauchamp[6,] Pell and Blossom[7] , Pickett[8] and Seiger[9.]

Beauchamp considered the processes at the counter electrodes and changed it from oxygen evolution to that of nitrite ion oxidation by inclusion of this ion at significant concentrations. This change was used to stabilize the pH of the mother liquor. He operated the bath at the boiling point of the mother liquor.

MacArthur proved that the nitrate ion reduction was indeed responsible for the pH changes that resulted in $Ni(OH)_2$ precipitation. When he buffered the solution to prevent pH change there was no precipitation.

Pell and Blossom operated their cathodic deposition at 85°C and used a constant voltage across the electrodes of the elec-trodeposition bath.

Pickett operated a 50% ethanol-water solution for the solvent of the nickel nitrate solution and worked near the boiling point of the solution, about 79°C. Part of the plaque-plate relationship is that the sintered material be clean and pure enough to be penetrated by the mother liquor. The alcoholic solution behaved as a wetting agent so that the mother liquor penetrates plaque material that would otherwise be unsuitable. The alcohol should either be pure ethanol or one in which the denaturant is not electroactive. Some denatured alcohol formulae interfere with the electrochemical precipitation processes.

a. Mechanism of the Electrochemical Process

The mechanism of any electrode process depends upon a number of controllable factors. These are the potential, the temperature, the substrate, the medium, pH and upon specific catalytic effects. Nitrate reduction was found by Tafel many years ago to result in production of hydroxylamine or ammonia depending upon whether the cathode material was mercury or copper. There are some differences in the mechanism of nitrate reduction depending upon whether the mother liquor is a $Cd(NO_3)_2$ or a $Ni(NO_3)^2$ solution. The primary nitrate reduction process for nickel nitrate depends upon current density/potential.

In the case of an acidified nickel nitrate solution the primary electrochemical reduction process may be written as:

$$NO_3^- + (6-x)H_2O + xH^+ + 8e^- \rightarrow NH_3 + (9-x)OH^-$$

This process occurs at all electrode/electrolyte interfaces so that more of the process takes place within the confines of the porous structure than on its geometric surface. If the acidified solution flows by the electrode fast enough then the pH can change at a significant rate only within the porous structure while the pH of the mother liquor changes at a much slower rate depending upon the solution flow rate and its total volume. The increase of local alkalinity within the pores causes precipitation of $Ni(OH)_2$ when the pH reaches 6.4, and continues the precipitation until the process is quantitative at pH 8.4. Thus, each unit of void volume now contains $Ni(OH)_2$ in an equivalent amount to the initial concentration of the solution. The precipitation step is represented as:

$$(9-x)Ni^{++} + 2(9-x)OH^- \rightarrow (9-x)Ni(OH)_2$$

The ammonia produced in the first equation is not gaseous, but is solubilized as NH_4NO_3; writing it as NH_3 is a simplification that helps understanding. The residual liquid within the pores continues to produce NH_4^+ and OH- ions until the pH is sufficiently alkaline, 13.5, to cause the ammonium ions to gasify. The gas production causes an expulsion of the exhausted solution from the pores. When the gaseous NH_3 contacts the acidified solution it dissolves again resulting in a partial a vacuum that draws in fresh mother liquor and the entire process begins again. This expulsion by gaseous ammonia is likened to the ammonia vapor fountain experiment[10] which demonstrates the solubility of gaseous NH_3 in water which is 28 volumes of gas to one of water.

The continued pH change of the first equation alters the coefficient x because a partial electrolysis of water occurs in the more basic medium:

$$8H_2O + 8e^- \rightarrow 4H_2 + 8OH^- \ .$$

The rate of pH change from 8.4 where precipitation is considered complete to pH 13.5 is calculated using the above equation. At a typical process current density of 0.08 A/cm², it requires 27 seconds to bring about the necessary pH change to gasify the NH_3 and purge the exhausted solution from the internal voids. The 27 second interval is a nonproductive period for metal hydroxide precipitation and affects the process efficiency.

Once a fresh aliquot of mother liquor enters the pores the electrochemical process continues resulting in an increased loading of the voids with active material. The $Ni(OH)_2$ (usually containing a small amount of Co^{++} ion which coprecipitates) deposits as a rather smooth coating on the metallic Ni particles of the sinter. Each time the exhausted solution is replenished the residual pore volume had be decreased by the volume of the deposited material. This material is black in color, not the typical green of $Ni(OH)_2$, suggesting a conductive, defect structure for the deposited $Ni(OH)_2$.

As the process continues and the weight gain builds up, the value of x, the stoichiometric coefficient, in the first equation can be calculated. A loading curve for an impregnation is given in Figure XIV-1.

The weight gain is readily converted to equivalents of $Ni(OH)_2$, and compared to the charge, expressed in faradays, used for the electrodeposition. The initial portion of the impregnation curve is used to calculate the slope M', and from the first two equations of this section the slope is given by:

$$M' = \frac{(9-x)/2}{8} .$$

This slope is not corrected for the non-productive intervals. For an actual impregnation it is estimated that the first ejection of spent electrolyte occurs in 770 seconds, but that 27 seconds are nonproductive for the second equation. If this factor, 0.96, multiplies the above equation then an expected loading slope, M, may be defined as:

$$M = 0.96 \frac{(9-x)}{16}$$

This last equation enables one to determine the stoichiometric coefficient experimentally.

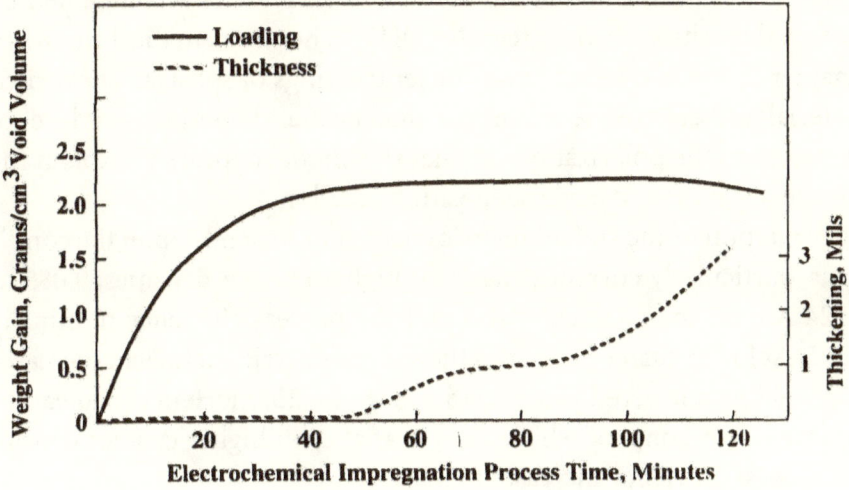

Figure XIV-1. Loading Curve and Thickness Changes During Electrochemical Impregnation.

The data in Figure XIV-1 indicates a saturation of the impregnation process as well as an electrode thickening as the saturation level is approached. The thickening occurs at a rate that maintains the saturated loading level which is about 2 g/cm³ of voids. Hence, the void volume of a thickened electrode is greater that its initial value due to stretching. These impregnations are carried out using constant current and the potential between the anode and cathode can be monitored as shown in Figure XIV-2. The anode in this case is inert, platinum plated upon tungsten, so that the anode process is oxygen production. When adding the anode and cathode primary processes, the following is obtained:

$2HNO_3 + H_2O \rightarrow NH_4NO_3 + O_2$,

the local pH becomes less acid and oxygen is produced at the inert anode. The pH in the process used for Figure XIV-2 was maintained by external means. The voltage across the electrodes is in the neighborhood of 0.7 to 0.8 volts until the fully loaded condition is approached when it increases to about 1.8 volts. This increase is accompanied by surface deposition of the green $Ni(OH)_2$. The mathematical model in Chapter II which is based upon concentric rings of substrate, deposited material and electrolyte relates the thickness and conductivity level of the layers to the polarizations so that this model appears to explain the voltage behavior and electrodeposition location.

Evaluation of the stoichiometric coefficient depends upon the conditions, particularly current density. At higher current densities, 0.085 to 0.124 A/cm^3, the coefficient is 3 and the process efficiency nominally 0.375. At lower current densities the stoichiometric coefficient increases to 5 yielding a lowered process efficiency of 0.25. McHenry[4] had earlier reported data, some of which show that at even higher current density the process efficiency increases to 0.5.

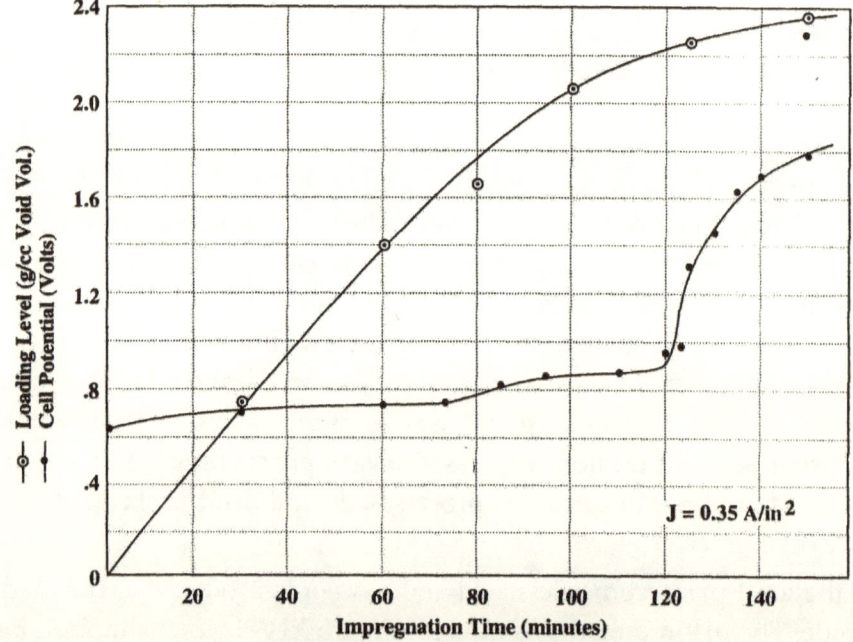

Loading Level[+] vs. Impregnation Time
Cell Potential vs. Impregnation Time
+Plaque Characteristics: Sinter Porosity = 78%; Plaque Thickness 34 mils

Figure XIV-2. Loading Curve and Changes in Potential During Constant Current Impregnation.

Having seen values for x = 1,3 and 5, it was speculated that the coefficient could increase to 7 at even lower current density and an experiment was successfully run to verify this expectation.

The conditions for impregnation by electrochemical deposition recommended are 0.08 to 0.124 A/cm^2 where electrodes below 1 mm thick can be impregnated within an hour. Flowby conditions also aid in keeping the surface free of deposits so that $Ni(OH)_2$ particles do not fall off the electrodes. The amount of $Ni(OH)_2$ deposited in these short time periods correspond to 7 to 8 times the material contained in the solution. A loading level of 1.65 g/cm~ corresponds to 18 meq/cm^3. The solution has an initial concentration of 2 to 3 meq/cm^3 (2 to 3M). Replenishment at these rates can not include a diffusion process, unless it is as a minor and negligible participant. The ammonia fountain as a replenishment scheme is consistent with the speed required and is a convective transport process. The ammonia production,its odor, and the bluing of wet litmus paper by electrodes freshly removed from the electrochemical impregnation bath are also consistent.

b. Counter Electrode Mechanisms

The electrochemical impregnation process has some dependency upon the counter electrode. The impregnating solution generally contains some cobalt, 2 to 10% of the overall concentration of metal ions. The process depletes Ni and Co, alters pH and increases the ammonia content. Some control of pH is necessary frequently or the solution becomes glutted with the basic salts. Several means have been used to maintain the solution.

Beauchamp[6] as mentioned earlier, incorporated nitrite ions in the electrolyte and used the inert anode. Under these circumstances the NO_2^- was oxidized to NO_3^- at the anode so that the oxygen generation potential was not encountered and, at the proper concentrations of nitrite, stabilized the pH.

The inert electrodes results in acid formation accompanying oxygen evolution and neutralization is necessary to avoid interference with the impregnation process.

When an active counter electrode is used such as nickel, it oxidizes and results in an increasing Ni^{++} ion content. Accompanying this dissolution of Ni the mother liquor requires acidification and a separate Co^{++} ion addition to compensate for the ratio change of Ni^{++} to Co^{++}. Krogeril used an active anode containing a NiCo alloy of such composition as to maintain the initial Ni to Co ratio. Seiger[12] used an inert anode and incorporated a second compartment. The second compartment contained a permeable bag containing $Ni(OH)^2$ and $Co(OH)^2$ so that the metal ions were in the same ratio as the bath liquid. In this case pH was maintained constant by reaction with these basic hydroxides and there was some replenishment of the Ni and Co ions in the proper ratio, although the replenishment was less than perfect. The increasing ammonium ion content of the solution was not affected.

The chemistries of these processes indicate the problems and procedures. Two examples should suffice, the first is the use of an active anode. In the following we let Ni* represent the mixed Ni and Co required by the bath.

$4Ni^* .-\sim 4Ni^{*++} +8e-$

while the net cathode process is the sum of the first two equations in this chapter:

$$[(9-x)/2]Ni^{*++} +(6-x)H_2O+xH^+ +8e^- \rightarrow NH_3 +[(9-x)/2]Ni^*(OH)_2$$

in neutral form, the combination is:

$$(x-1)HNO_3 +(6-x)H_2O+Ni^* \rightarrow NH_4NO_3 +[(9-2)/2]Ni^*(OH_2)+[(x-1)/2]Ni^*NO_3$$

Regardless of the stoichiometric coefficient acid is consumed and must be replaced. The stoichiometric coefficient governs whether the nickel preferentially deposits as $Ni^*(OH)^2$ within the pores or whether the Ni^{++} concentration of the bath increases.

The second example is the use of an inert anode. In this case the anode process is the electrolysis resulting in oxygen evolution:

$$4H_2O \rightarrow 8H^+ + 2O_2 + 8e^-$$

When this equation is combined with the sum of the first two equations, the result is:

$$(10-x)H_2O + [(9-x)/2]Ni*(NO_3)_2 \rightarrow NH_3 + [(9-x)/2]Ni*(OH)_2 + 2O_2$$

which causes a depletion of the Ni^{*++} in the bath and an acidification. However, by neutralizing the acid formed with a mixture of Ni and Co bases, such as $Ni(OH)_2$ and CoO (or $Co(OH)_2$) the pH is stabilized and most of the cations replaced. The neutralization step is:

$$[(8-x)/2]Ni*(OH)_2 + (8-x)H^+ \rightarrow [(8-x)2]Ni^{*++} + (8-x)H_2O$$

resulting in the overall reaction:

$$0.5Ni*(NO_3)_2 + 2H_2O + [(8-x)/2]Ni*(OH)_2 \rightarrow NH_3 + [(9-x)/2]Ni*(OH)_2 + 2O_2$$

According to this last equation the rate of Ni^{++} and Co^{++} ion depletion is markedly decreased. In practice, instead of bath adjustments on a daily basis, it is done on alternate weeks. The water adjustment is merely a bath level adjustment. The bags containing the mixed Ni and Co bases have to be replenished occasionally as needed. The buildup of ammonium salts occurs in all instances. Hausler[3] observed that there is no interference as long as the concentration remains below 2 molar.

No clever way has been found to decrease or control the ammonium ion concentration. A large bath volume to pore volume ratio is used so that the molar rate at which ammonium ion does increase is not very great. When the concentration approaches the limit the solution may be replaced and the ammonium ion contaminated solution made basic to precipitate the mixed Ni and Co hydroxides. After washing and drying these can be used in the neutralization bags.

An experimental tank for use in electrochemical impregnation is shown in Figure XIV-3[13.] There are two sections. One section is used for the impregnation process and the porous plaques are immersed between counter electrodes of platinum-plated titanium. The counter electrodes are adjustable in depth since the anodes must be a little smaller than the work pieces to assure uniform impregnation. There is a manifold at the bottom of the impregnation section through which the solution is transported from the reservoir and distributed over the faces of the work pieces.

There is a weir over which the effluent from the impregnation section reaches the reservoir. The cloth bags with the solid-mixed Ni and Co hydroxides are in this compartment. The bath is maintained at about 85°C by immersion heaters and a pump is used for mixing, stirring and transporting the solution to the manifold. The solution between the two sections is filtered. Finally. there is a triangular cover used to prevent evaporation by its reflux action.

The work pieces are mounted on the bus bars externally. From the moment the sintered nickel contacts the hot acidified nitrate bath corrosion of the plaques start. This corrosion could be halted either by using a cadmium wire for cathodic protection as a sacrificial material or entering the solution with current on. If Cd is used, it then behaves as an antipolar matter and a contaminant that soon becomes undesirable. With current on there is a danger of arcing. Alternatively, if the contact time before current is turned on is minimized, then corrosion can be held to less than 2%[14] Contrasting this to aerospace electrodes where 18 to 40% corrosion has been observed this amount is acceptable.

c Influence of Plaque Porosity on Nickel Oxide Electrodes.

There is an effect of plaque porosity upon utilization of nickel hydroxide deposited. Most utilization data had been based upon a one faraday per mole process:

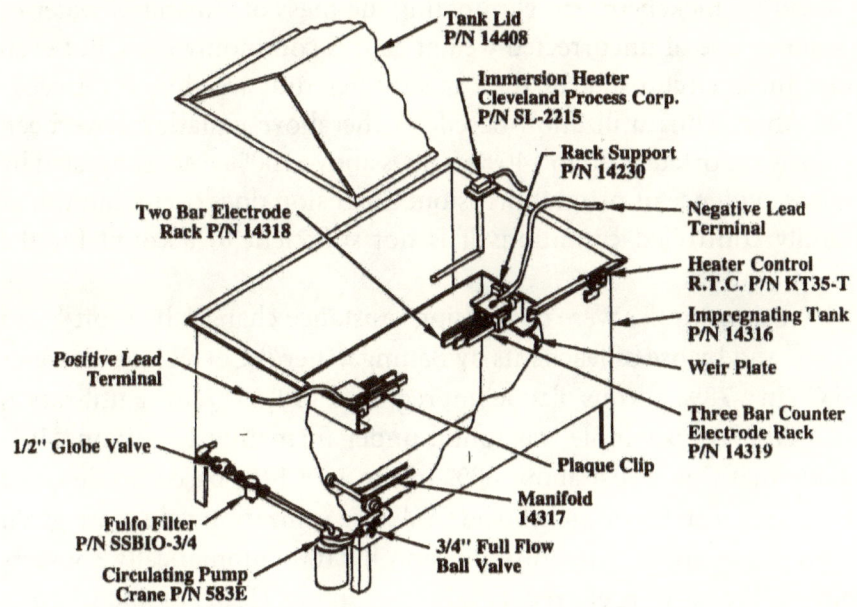

Figure XIV-3. Diagram of An Impregnation Tank with Two Sections. One Section for Impregnation and the Second for SeLf-neutralization and Replenishment.

$$Ni(OH)_2 + OH^- \rightarrow NiOOH + H_2O + e^-$$, but there is evidence that the process is more complex. Jackowitz and Feldman[15] presented spectroscopic evidence for a nickel oxidation state of 3.86 and more recent work by Cornielsen[16] and Corrigan[17] are in agreement. Tichenor[18] provided experimental evidence that Ni is not readily reduced below an oxidation state of 2.12. The theoretical capacity of the nickel oxide electrode should be based upon 1.74 F/M and not upon one F/M as indicated in the above equation. Tichenor suggests that below an oxidation state of 2.12 the defects of the crystals are eliminated and the material becomes an insulator, while Cornielsen has a similar explanation for the limitation of the upper oxidation state. The weight gain should

be based on nickel content eliminating the mass of Co and of water of hydration. Use of uncorrected weight gain is commonly used, but even under these circumstances the capacity per unit weight gain exceeds 0.288 Ah/g. Thus utilization based on the above equation have been variously reported at up to 140%. Values above 100% were suggested by some as evidence of plaque corrosion. Corrosion does occur, but under carefully controlled conditions it is not sufficient to account for the high utilization.

Corrosion studies were done using resistance changes by Scott[19] and Seiger[15] and by use of leachants by Baumgartner[20]. Correcting for water and Co in a 78% porous plaque impregnated to 1.65 g/cm[3], a utilization of 1.4 faradays per mole was found under formation conditions. It is notable that this is just about 80% of the 1.74 F/M based on limits of the oxidation states of nickel oxides. Where utilization values are given in percentage in the literature division by 100 automatically converts them to the now preferred faraday per mole (F/M) nomenclature. Reporting values in excess of 100% is esthetically unappealing and is avoided by changing to a faraday per mole basis.

The first observation of the interrelationship of plaque porosity and utilization was reported to the Air Force[21] and reproduced here as Figure XIV-3. These data were based upon historical formation processes. Formation is a term used in the industry for improvement of utilization of active materials and came to the nickel electrode field via the lead acid battery industry. Subsequent to this early work the formation process was modified to incorporate the 1.74 F/M now expected, and the fact that charging of the nickel oxide electrodes are inefficient due to parasitic oxygen evolution[22]. The formation recipe was changed to a 2F/M charge in over 2 hour period. As a result of the high rate, high capacity charge utilization in a 79% porous sinter was increased from 1.1 F/M to 1.4 F/M.

The data of Figure XIV-4 were used to compute a maximization of capacity per unit volume of the sintered body. The optimum, porosity is

77% but variation from this value is readily tolerated as shown in Table XIV-1.

Table XIV-1
Capacity Density for Selected Sinter Porosities

Sinter Porosity, %	Capacity, Ah/in³ Sinter
77	7.79
70	7.7
85	7.68

The capacity per unit weight is also determined from the data of Figure XIV-1, but the mass associated with the grid has an influence as well as the thickness of the electrode. Figure XIV-5 illustrates how the gravimetric capacity depends upon porosity and the thickness of the electrode. However, as the electrodes are made thicker current densities are consequently increased which affect the degree of polarization. Plate thickness and performance requirements must be considered in battery, cell and electrode design.

Loading levels are a factor in subsequent swelling. The fact that the nickel oxide deposits are on the internal surfaces of the sintered nickel, and are apparently well adhered gives rise to hoop stresses. The

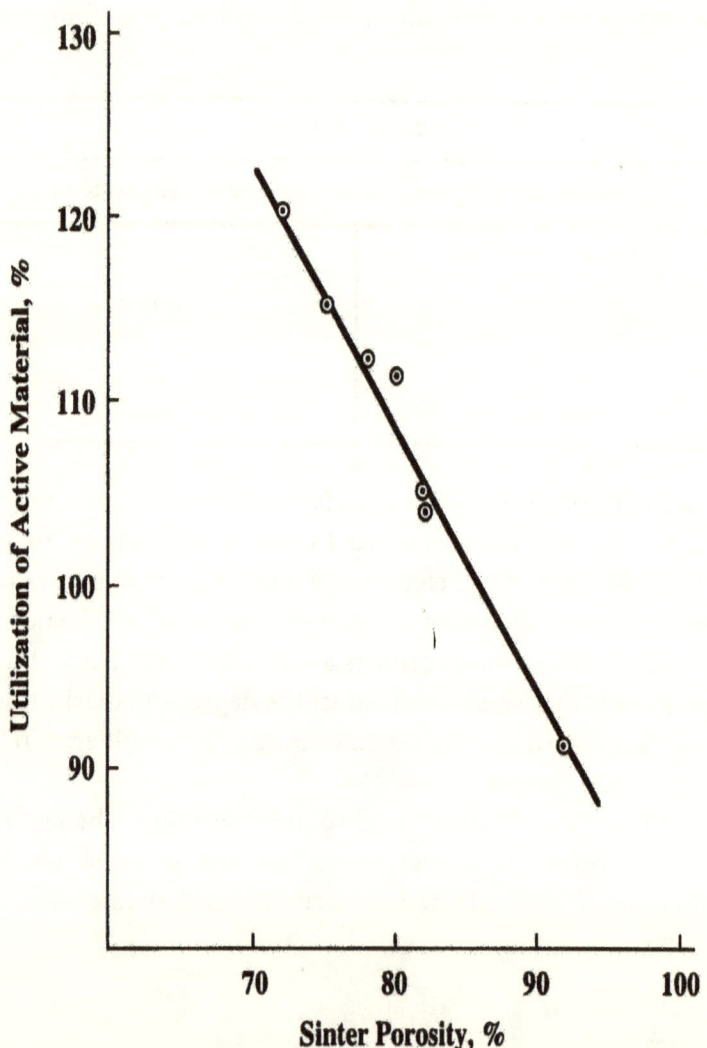

Figure XIV-4. Nickel Oxide Utilization Dependence Upon Sinter Porosity.

stoichiometric and density changes of the active material result in volume changes of the nickel oxides. These hoop stresses eventually cause fatigue of the fused joints between sintered particles. Studying the loading level and utilization along with the swelling rate of electrodes led to a selection of 1.65 g/cm^3 voids. Such electrodes show a reduced swelling by with useful life increased by a factor of 4 with improved volumetric and gravimetric capacity densities.

There are problems associated with less than fully loaded electrodes. They tend to be less uniformly impregnated. Three steps have been taken to improve uniformity. The first is use clean plaque for impregnation. There is a test for cleanliness based on capillarity. When held upright in a trough of water a plaque sample should show a capillary rise in which the leading edge of the wetness is a straight line. Failing this test does not make the plaque unusable per se, since it can be cleansed by treatment in a reducing atmosphere at temperatures which are high but (about 700°C) not result in oversintering. This furnace treatment can be done rapidly. Alternatively, the plaque may be cleaned by a cathodization in alkali for just a few minutes at currents just great enough to cause free flow of hydrogen gas.

The second precaution for obtaining uniformly impregnated electrodes at less than full loading levels is the use of undersized anodes. Finally, the third contributor to uniformity is the use of moderate current densities that produce the stoichiometric coefficient of value 3 in the first equation.

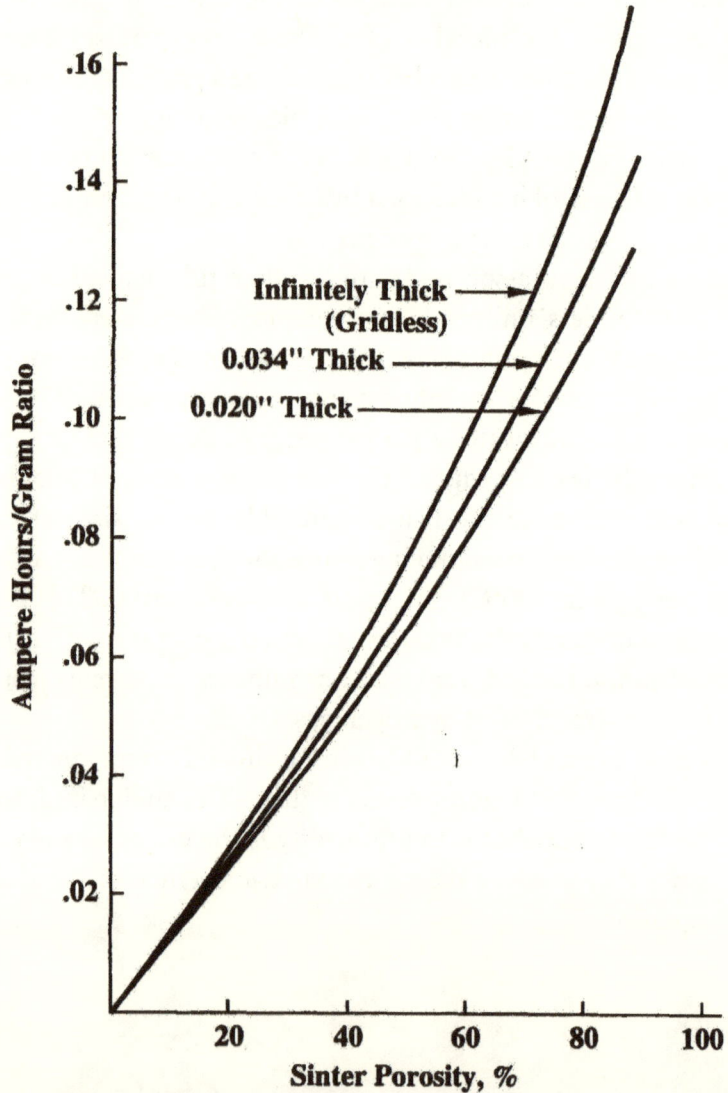

Figure XIV-5. Gravimetric Capacity Dependency Upon Porosity and Thickness of Sintered Plaque.

1   A. Fleischer, J. Electrochem. Soc, 94,289 (1948)

2   L. Kandler, German Patent 1,133,442, July 1962.

3   E. Ha~sler, Fifth International Power Sources Symp., Sept. 1966, Brighton, England.

4   E.J. McHenry, J. Electrochem. Technology, 5, 275 (1967)·

5   D. M. MacArthur, Power Sources Symp.. 1970, Brighton, England.

6   R.L. Beauchamp, U.S. Patent 3,653,967 April 1972.

7   M.B. Pell and R.W. Blossom, U.S. Patent 3,507, 690 Apil 1970.

8   D.F. Pickett, U.S. Patent 3,827,911 August 1974.

9   H.N. Seiger, U.S. Patent 4,292,143 September 1981.

10 F.A. Cotton, *Advanced Inorganic Chemistry*, Interscience Puablishers, New York, 1966.

11 H.H. Kroger et.al., U.S. Patent 3,979,223, Sept. 1976

12 H.N. Seiger and V.J. Puglisi, U.S. Patent 4,120,757 October, 1978.

13 H. N. Seiger, Proc. Power Sources Symp., 1976.

14 H.N. Seiger, U.S. Patent 4,292,143, Sept. 1981.

15 J.F. Jackowitz and D.W. Feldman, ECS Meeting. Miami, Oct. 1972.

16 B.C. Cornilsen, J. Power Sources, *22*, 351, (1988)

17 D. Corrigan, J. Electrochem. Soc. 136, 613 (1989).

18 R.L.Tichenor, Indust.& Eng. Chem, *44,973* (1952)

19 W.R. Scott, TRW Report,"A Study of Degradation of Plates for Nickel Cadmium Spacecraft Cells" JPL Contract 953649

20 C.E. Baumgartner, J. Electrochem. Soc., *135*, 36 (1988)

21 H.N. Seiger et. al.,Technical Report AFAPL-TR-74-56, Part I. August, 1974.

22 H.N. Seiger and S. Lerner, Proc. Intersoc. Energy Conv. Eng. page 353, 1967.

XV

Impregnation with Cadmium Compounds.

The nickel plaque impregnated with $Cd(OH)_2$ serves as the negative electrode in the secondary Ni/Cd battery cell. In this form the electrode is capable of high rates of charge and discharge, and is also capable of reacting with oxygen under certain conditions which are designed into sealed cells. The nickel plaque is sintered from Ni powders having an acicular structure with dimensions of 2 to 3.5 μ. Depending upon the actual particle size and sintering conditions the porosity of the sintered material can range from about 65% to 90%. More usually the porosities are 78% to 86%. The pore sizes average 10 to 17 μ[1].

The impregnation of these pores with $Cd(OH)_2$ in industry is commonly done as described by Fleischer[2,] or some variation of that vacuum immersion method. The plaque is immersed in a solution of a salt such as $Cd(NO_3)_2$ with the aid of a vacuum. The plaque is dried, the salt converted to $Cd(OH)_2$, washed free of alkali and redried. The most concentrated salt solution is the fused salt which corresponds to about a 7 M concentration. Since a desirable impregnation level is about 16 mmoles of $Cd(OH)_2/cm^3$ of void volume, several impregnation cycles are necessary. The actual number depends upon the solution concentration, pore penetration and the idiosyncrasies of the process. It had been observed[3] that while the $Cd(NO_3)_2$ may be uniformly distributed in the voids, the resultant $Cd(OH)_2$ is in a layer just under the surface. Subsequent layers continue to build inward toward the center of the plaque. Usual commercial processes require 4 to 9 vacuum impregnation cycles. The process is costly because it is labor intensive. Aside from this disadvantage, corrosion of the sintered material by the $Cd(NO_3)_2$ is encountered; bubbles due to gaseous corrosion products no doubt affect the pore penetration. The alternative to vacuum impregnation process for deposition of $Cd(OH)_2$ into the voids is electrochemical impregnation. The process and the conditions have undergone considerable change since first described by Kandler[4] in 1962, and as practiced with the work about to be described. The changes have been in temperature, pH current density, concentration, and uniformity of current distribution. Commercial interest in electrochemical impregnation increased as these changes were brought about because:

1. High loading levels are achieved with a single impregnation cycle
2. Utilization of the active materials is higher than previously observed.
3. The process is uniform and controllable which enhances repeatability.
4. Corrosion of the plaque is virtually eliminated.
5. The process is cost effective.

The variations in parameters were made in spite of the fact that the chemistry and mechanisms were not well understood.

Kandler[4] hypothesized that the NO_3^- ion was reduced to NH_4^+ through a number of consecutive reactions. Hausler[6] confirmed the NH_4^+ production in nearly quantitative amounts. Hausler[6] now recognizes that some of the NH_4^+ or NH_3 was present in the solution "dragged out" with plaque removal. These workers believed that a local pH change brought about by the ammonia production resulted in the precipitation of the $Cd(OH)_2$.

It must be noted that there have been no studies of the precipitation of $Cd(OH)_2$. Rather past investigations have been on the precipitation of $Ni(OH)_2$ and logical argument by analogy used for $Cd(OH)_2$ precipitation. Investigating the $Ni(OH)_2$ reaction, MacArthur noted the pH change accompanying NO_3^- ion reduction, but neglected the confirmed finding of Kandler and Hausler. As a consequence, the work was not definitive. Beauchamp[7] recognized the uniqueness of the NO_3^- ion, because of its reducibility under the conditions involved.

Pickett[8] established by analysis that the material in the electrodeposit is a mixture of Cd and $Cd(OH)_2$. Maurer[9] noted that the functional dependence of weight gain on process time appeared to be a superposition of a linear and a non-linear function. Most of the modern electrochemical impregnation processes utilize $Cd(NO_3)_2$ solution of 2 to 3 M concentration, yet deposit 10 to 18 mmoles of $Cd(OH)_2/cm^3$ of void volume. This is an augmentation factor of about 10 which occurs in period of time ranging from 10 minutes to about 1 hour. It can be easily shown that diffusion can not support the necessary flux for this type of augmentation. Further, if diffusion processes played a predominant role, then some sort of deposition gradient is expected within the plaque. That is, the interior would be relatively free of Cd species, while the pores near the surface become blocked. The initial deposit is uniformly distributed throughout the porous plaque.

A subsequent paper[10] in which the electrochemical impregnation of $Ni(OH)_2$ was investigated proposed a mechanism for that process. It was suggested that the NO_3^- ion was reduced to NH_3. The complex between Ni^{++} and NH_3 does not follow Lamberts Law: the absorption peak shifts to shorter wave lengths as the ammonia concentration is increased. Hence, NH_3 can not produce a sufficient concentration of OH^- ions to cause precipitation of $Ni(OH)_2$. To account for the precipitation, the NO_3^- reduction process was written as:

$$NO_3^- + (6-x)H_2O + xH^+ \xrightarrow{j_e} NH_3 + (9-x)OH^-$$. Such an equation expresses the knowledge we have, and becomes more useful as a diagnostic tool since there is a relationship between charge and the amount of OH^- ions produced. The subsequent precipitation reaction:

$$\frac{9-x}{2} Ni^{++} + (9-x)OH^- \xrightarrow{k} \frac{(9-x)}{2} Ni(OH)_2$$

gives rise to an F/M ratio from which the value of x depends on the slope of the weight gain of $Ni(OH)_2$ electrodeposited versus time. Such curves are readily determined by experiment. $Ni(OH)_2$ is not reducible under the conditions existing within the plaque while $Cd(OH)_2$ is capable of being reduced to elemental Cd. Hence the mechanism previously proposed for the electrodeposition of $Ni(OH)_2$ can be modified for the electrochemical impregnation of Cd and $Cd(OH)_2$ into porous sintered nickel plaque. The modifications are (1) that the coefficient x in the above equations must be determined for the different cation involved and (2) the assumption that $Cd(OH)_2$ reduction to Cd is competitive with the precipitation reaction.

a. Theory

The cathodization of the porous plaque in a slightly acidified $Cd(NO_3)_2$ solution at current densities ranging from about 0.08 A/in^2 to 0.32 A/in^2 results in the reduction of NO_3^-. The production of OH$^-$ ions within the pores increases the local pH. When the pH reaches a value of 6.8 $Cd(OH)_2$ begins to precipitate, and the precipitation is considered quantitatively complete at pH 8.8; that is, the concentration of Cd^{++} ions is decreased from 2 M to 2mM within the pores. The local pH within the voids then continues to increase until at a value of 13.5 the NH$_3$ vapor pressure reaches 1 atmosphere. The vapor pressure causes the exhausted solution to be expelled from the pore. When the NH$_3$ vapor contacts the acidified solution sweeping along the surface, it dissolves again which causes a vacuum inside the pores that draws in fresh solution so that the process can repeat.

The actual void volume is decreased slightly since solid $Cd(OH)_2$ is present. The amount of solution now present in the pore due to this replenishment process is correspondingly less than the initial condition.

Thus, from a 2 M solution 2 mmoles of $Cd(OH)_2$ are deposited per unit volume initially; 1.88 mmoles of $Cd(OH)_2$ deposited after the replenishment, and a similarly decreasing quantity deposited after each further replenishment. After 10 such replenishment processes the weight gain per unit of initial void volume as $Cd(OH)_2$ is 16 mmoles or 2.38 g.

The mechanism thus far accounts for a uniform impregnation within the pores, and a higher loading level achieved without diffusion playing a dominant role. The reduction of $Cd(OH)_2$ may now be considered.

While the current J is passing through the cathodes, some of the $Cd(OH)_2$ is reduced to Cd, and this reaction is in competition with the reduction of NO_3^- to NH_3:

$$Cd(OH)_2 + 2e^- \xrightarrow{j_p} Cd + 2OH^-$$

It is reasonable to assume that no other reactions are occurring so that the total current density J is equal to the sum of the NO_3^- reduction current j_o and the parasitic $Cd(OH)_2$ reduction current j_p:

$$J = j_o + j_p$$

where j_o and j_p each vary with time. Recognizing that the precipitation reaction analogous to the equation for Ni^{++} is:

$$\frac{9-x}{2}Cd + (9-x)OH^- \xrightarrow{k} Cd(OH)_2$$

and that the sum of these equations is:

$$NO_3^- + (6-x)H_2O + xH^+ + \frac{9-x}{2}Cd^{++} + 8e^- \rightarrow NH_3 + \frac{9-x}{2}Cd(OH)_2 .$$

The number of faradays of charge, z_1, associated with the precipitation of one mole of $Cd(OH)_2$ is defined as:

$$z_1 = \frac{16}{9-x}$$

If the number of moles of $Cd(OH)_2$ is denoted by n, then the rate at which $Cd(OH)_2$ is precipitated is given as:

$$\frac{dn}{dt}\Big|_f = \frac{j_c}{z_1 F}$$ where F is the faraday constant.

The rate at which $Cd(OH)_2$ is reduced according to the appropriate equation above is given by:

$$-\frac{dn}{dt}\Big|_r = \frac{j_p}{z_2 F} = \Theta J \frac{n}{v_o}$$

where Θ is an adjustable parameter, v^o the void volume per unit area of porous plaque and z_2 has a value of 2 since 2 faradays of electricity are required to reduce one mole of $Cd(OH)_2$ to Cd. If the plate is ana-lyzed for Cd and $Cd(OH)_2$ content after the plate is impregnation, destroyed and no additional information may be obtained; alterna-tively, plates may be discharged after appropriate treatment so that the only species present may be assumed to be $Cd(OH)_2$. Thus, the number of moles of $Cd(OH)_2$ under these particular conditions may be repre-sented as n_f. These rate equations may be substituted into the constant current equation yielding:

$$J = z_2 J F \frac{n}{v_o} + z_1 F \frac{dn}{dt}\Big|_f$$

Separating variables, integrating and using the boundary condition that n=O at t=O, the following equation is obtained:

$$\frac{n}{v_o} = \frac{1}{z_2 F \Theta}\left[1 - \exp\left(-\frac{z_2}{z_1}\Theta J \frac{t}{v_o}\right)\right]$$

When both sides of the above equation are multiplied by the mo-lecular weight of $Cd(OH)_2$, M, and noting that nM/v_0 is the loading level, L, in grams $Cd(OH)_2$ per unit volume of initial pore volume, the following equation is obtained:

$$L = L_o\left[1 - \exp\left(-\frac{z_1}{z_2}\Theta J \frac{t}{v_o}\right)\right]$$

where $L_o = \dfrac{M}{z_2 F \Theta}$. While the above equation requires the loading level rate to steadily decrease toward zero which is necessary condition for the relationship, a better form for test purposes is:

$$\ln\left(1 - \frac{L}{L_o}\right) = -\frac{z_2}{z_1} \Theta J \frac{t}{v_o}$$

so that a semilogarithmic plot is used for data analysis.

At the start of the electrochemical impregnation process n_f has a small value so that the parasitic current my be neglected. The initial slope of a loading curve is used to estimate x. Under these initial conditions J may be substituted for j_o.

b. Experimental Findings

Plaque was sintered from INCO types 287 and 255 powders manufactured by the Mond process. Thickness ranged from 0.6 mm up to 1.0 mm. The impregnation bath used 2M $Cd(NO_3)_2$ maintained near the boiling point. The current density was based on the height and width of the plaque although the counter electrodes were on either side of the plaque: this procedure for defining current density was consistent with Pickett[6] but differs from the convention used by MacArthur[5]. The hot impregnation solution was pumped from a reservoir through a manifold on the bottom of the impregnation tank. In this way the solution traveled up and along the faces of the cathode resulting in tangential flow.

To minimize corrosion of the plaque by the solution, current was turned on within one minute of contact. Separate experiments indicate that corrosion could be between 0 and 1%, depending upon the fastidiousness of the worker. Data obtained in the laboratory have essentially no corrosion, while data obtained in the manufacturing facility involves corrosion of about 1% of the nickel plaque. Such findings are consistent with Scott's[11] data measured by electrical resistance of plaque and plate.

Electrodeposition was carried out at the indicated current density for various time periods. The test plates were quickly removed, washed with water and then polarized in an aqueous solution of KOH. The washing was performed to remove NO_3^- ions which interfere with the cathodization of $Cd(OH)_2$. The plates were scrubbed, washed to pH 8 and dried in a convection oven in which entering air was lead through a NaOH trap to remove CO_2. Following this they were made into test cells and charged/discharged several times. The capacities were determined. After determining capacity, the discharged electrodes were washed and dried.

It is at this point that the weight gains were determined; measurements were taken that established there was no thickening of the electrodes. The utilzations were based on weight gain and the measured capacity. Sample plates were subjected to electron microscopy. This technique for studying the plates microscopically was to freeze by immersing in liquid nitrogen and then cracking the plates by bending. In this way the location and relative amounts of material ranging from near the surface to near the grid could be estimated.

The first situation studied using scanning electron micrographs was that of the impregnated electrode that was oxidized so that no elemental Cd was present; all of the cadmium species being present as $Cd(OH)_2$. The material appeared to be the same throughout the electrode indicating a uniform deposition of the cadmium species. The crystals have a dimension of 1 to 2 μ.

SEM's of the same kind of electrodes after 4 charge-discharge cycles constituting the process termed formation were also studied. The crystals have grown to about 10 m and the material was redistributed. In the region close to the grid the active material crystals were tightly packed into the pores, and there was obvious twinning of the crystals.

The region midway between the grid and the surface had some $Cd(OH)_2$ crystals. While the crystals were also about 10 μ, there was less twinning giving the impression the material was less densely packed.

The region just under the surface showed few crystals of $Cd(OH)_2$. This region appears virtually unimpregnated. This kind of active material distribution is just opposite to that found with Fleischer type methods where most of the $Cd(OH)_2$ was near the surface, and the region near the grid was unimpregnated. The tight packing of the crystals make it more difficult to remove entrapped air from the negative electrodes, an important consideration in manufacturing of sealed cells.

A typical loading curve taken at 0.088 A/cm^2 into a plaque 85% sinter porosity is shown in Figure XV-1. The shape of the curve appears to be consistent with the loading equation derived above. Table XV-1 gives some experimental loading levels. The ratio of equivalents as $Cd(OH)_2$ deposited per faraday of electrolyzing current in the electrochemical impregnation bath for the shortest experimental times, being guided by the last comments in the theoretical section, are 1.28, 1.00, 1.14 and 1.08. From these values of F/E (faradays per equivalent) it is deduced that the value of x, defined in the impregnation mathematical equation, is equal to zero.

A plot of $\ln\left(1 - \dfrac{L}{L_o}\right)$ versus normalized time is shown in Figure XV-2. Normalization of time is obtained merely by dividing real time by void volume unit area. The slope of such curves appears to depend upon the characteristics of the nickel powder. Inco Types 255 and 287 powders both yield the required logarithmic relationship but differ from each other. The curve fit to the derived

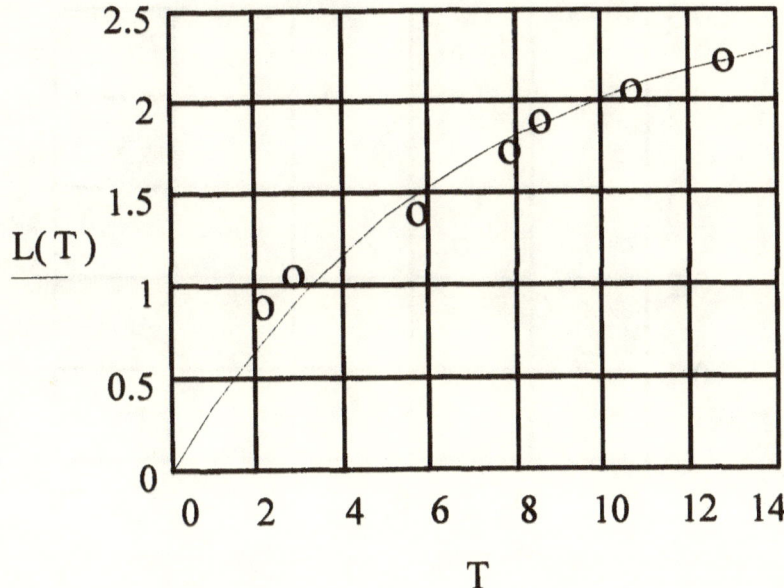

Figure XV-1. Loading Curve of $Cd/Cd(OH)_2$ into Porous Ni (as $Cd(OH)_2$).

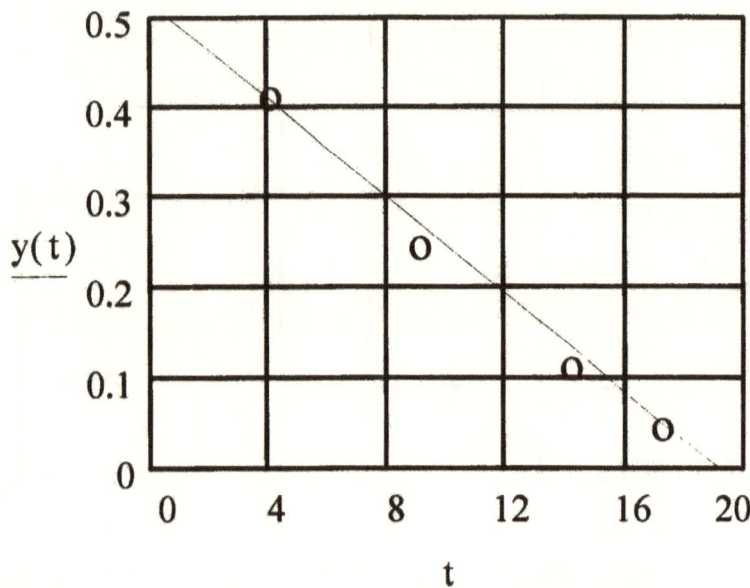

Figure XV-2. The Semi Log plot for the Loading Level Curve.

TABLE XV-1.	
Current Density, A/cm^2	Loading Level, grams Cd(OH)$_2$/cm^3 void volume
0.32	1.8
0.22	2.1
0.088	2.6

is equation generally satisfactory.

c. Discussion

Electrochemical impregnation by the described method results in a high loading level in a relatively short period of time. The deposition of 2.2 g/cm^3 of void volumes may be expressed as 30 mg. The solution from which the impregnation was made contains 4 mg/ml. A diffusion theory was considered and then rejected since diffusion rates (1) would not support the quantity of materials deposited and (2) a diffusion process is expected to deposit more material at the pore/solution interface than within the plaque, i.e. the deposited material would not be uniform in distribution within the plaque.

The mechanism given is consistent with the data obtained in several respects. The necessary reactants, NO_3^- and Cd^{++}, are uniformly present in the plaque when current flow is commenced. One mole of NO_3^- is reduced to NH_3 with simultaneous production of 9 moles of OH^- by passage of 8 faradays of cathodic current. The initial portions of the loading curves are consistent with this information. The NH_3 has been detected, and had been quantitatively established previously by Hausler. Our ammonia data are in concurrence. The critical electrode area must be the totality of the electrode/electrolyte interface which includes all the surface area within the plaque that is wetted by electrolyte. Hence, the interfacial area is that measured in double layer capacity work. The geometric surface represents about 1% or a little more, of the interfacial area. Since the volume of solution in the reservoir is large and flows over the plaque surface, the surface pH is nearly that of the reservoir solution. Hence, the amount of surface deposit is small. There is a local pH change within the pores. As the entrapped solution becomes more alkaline $Cd(OH)_2$ is precipitated. The initiation of precipitation is calculated to start at pH 6.8. When the precipitation is quantitatively complete, at about pH 8.8, the rate of pH change increases rapidly. When the local pH within the voids reaches 13.5 the ammonia vapor pressure reaches 1 atmosphere, and is responsible for expulsion of the exhausted electrolyte from the pore.

The exhausted solution is thus emptied into the reservoir and the ammonia dissolves in the acidic reservoir solution causing a vacuum within the pore. This is followed by a refilling of the pore with fresh solution. The process repeats itself. Since 4 mg were deposited by exhaustion of the initial quantity of Cd^{++}, the residual void volume is decreased to 0.97 cm^3, the total number of replenishment processes is 10 for an average time of one replenishment every six minutes.

Complicating the process, there is a reduction of electrodeposited $Cd(OH)_2$ to Cd. As the quantity of $Cd(OH)_2$ increases, the reduction rate increases. As a consequence, there is a premature limit to the quantity of Cd species impregnated and elemental Cd is present in the plates.

After formation the active material is redistributed so that there is more material close to the grid than is near the surface. The crystals of $Cd(OH)_2$ grow considerably during the formation process. In spite of the larger crystals the material is quite active yielding utilzations in excess of 90%. The heavy impregnation increases the mechanical strength of the plates, which is especially desirable with very porous plaque.

1   H.N. Seiger et.al., Paper presented at ECS Meeting, October, 1974.

2   A. Fleischer, Trans. Electrochem. Soc., *94*, 289 (1948).

3   H.N. Seiger, Unpublished 1972 data..

4   L. Kandler, Dechema Jahrestagung, 1962.

5   E. Hausler, Power Sources Sump. Brighton, 1966.

6   E. Hausler, private communication.

7   R.L. Beauchamp US Patent 3,653,967 April 1972, and US Patent 3,353,101 March 1971.

8   D. Pickett, NASA/GSFC Battery Workshop, 1973.

9   D. Maurer, NASA/GSFC Battery Workshop, 1973

10 H.N. Seiger & V. J. Puglisi, Power Sources Symp. Atlantic City, 1976.

11 W.R. Scott, A Study of Degradation of Plates for Nickel-Cadmium Spacecraft Cells, Final Report JPL Contract 953649, June 1974.

XVI

Molten Salt Impregnation

Sintered plate nickel cadmium battery electrodes have been impregnated with the active materials, nickel hydroxide and cadmium hydroxide using the method described by Fleischer[1.] Aqueous solutions of the nitrate salts were introduced using a vacuum assist. This was followed by drying to leave crystals of the hydrated nitrates which were then converted to the hydroxide using a base of aqueous sodium hydroxide or potassium hydroxide. The penetration of the base to the interior of the plaque was enhanced by cathodization.

Since the concentrations of the salts are in the range of 2 to 3 molar, the process must be repeated a number of times to achieve adequate loading levels and commercial viability. These repetitive steps are costly, labor intensive and time consuming, and it would be highly desirable to use more concentrated solutions and decrease the number of impregnation cycles to achieve targeted weight gains. The hydrated nitrate salts of both nickel and cadmium melt at about 60°C and the melt represents the most concentrated solutions. When these were tried experimentally each salt presented it own problems limiting success.

a. Difficulties Encountered Using Nickel Nitrate Melts

During the conversion of nickel nitrate to nickel hydroxide a flocculent form of the green hydroxide was formed. By the time the second impregnation was obtained all the plates blistered and were not usable. Some blistering was even encountered after the first impregnation in the conversion step.

Varying the nature of the base, sodium versus potassium hydroxide, and the concentrations were effective to a small extent in changing the flocculency and color or hue of the precipitate so that blistering was delayed from the first to the second cycle for some plates, but all were blistered by the time a third impregnation cycle was done. Since three impregnation cycles were calculated to be necessary for commercial viability, the improvements were inadequate.

The cause of the blistering and spalling appears to have been due to stresses set up by the hydrated hydroxides of nickel and a denser form of nickel hydroxide was needed to solve the problem. It was then reasoned that if the dried nickel nitrate was anodized instead of cathodized then the charged form of nickel oxide would be generated and this material is the densest form of the nickel oxides. This is the material present in the charged state of commercial positive plates of the nickel cadmium battery.

Plaque impregnated with the molten salt were anodized in an aqueous potassium hydroxide after the nickel nitrate salt froze. In this way levels were reached, without blisters, that were adequate. This method was not pursued because three impregnation cycles were needed and competitive electrochemical impregnation provide superior characteristics at comparative commercial costs.

b. Difficulty with Cadmium Nitrate Melts

Vacuum impregnation of sintered nickel plaque with molten cadmium nitrate occurs readily, but two problems were encountered. The first is supercooling of hydrated cadmium nitrate. This was overcome by rubbing the surface of the plaque with a stiff bristled brush. The frozen cadmium nitrate was then converted to cadmium hydroxide by cathodization in aqueous potassium hydroxide. The subsequent impregnation cycle resulted in very little addition weight gain.

Microscopic examination of the plaque cross section revealed a film of the precipitate on and just under the plaque surface. This dense film of cadmium hydroxide apparently acts as a membrane that permits the soluble salts to be leached during washing and prevents entry of the impregnating solution to the interior of the plaque.

In principle the interior of the plaque could be opened by destroying the membrane. This should be accomplished by reduction of cadmium hydroxide to elemental cadmium. An electrochemical reduction can not be carried out effectively in the presence of the nitrate ion. Reduction of nitrate occurs competitively with reduction of the Cd^{++} ion. Hence, after the initial cathodization to convert $Cd(NO_3)_2$ to $Cd(OH)_2$, the residual nitrate ion was removed by washing. When the nitrates are substantially removed, the partially impregnated plaque was recathodized. Washing and drying was done in the absence of an oxidizing atmosphere so that the elemental cadmium was not oxidized inadvertently. The entrance to the interior was recovered and the second impregnation with molten salt was readily accomplished. Calculations showed that the recathodization need be done only on the penultimate impregnation cycle. Only three impregnation cycles were needed to achieve the desired loading levels.

Cadmium hydroxide appears to be slightly soluble in the aqueous potassium hydroxide electrolyte. Examination of the cross section of the electrode after formation reveals a distribution of $Cd(OH)_2$ crystals was just opposite to the distribution found after the first cathodization on the initial molten salt impregnation.

Again, the molten salt impregnation became a less desirable method oflectrode manufacture than the electrochemical method.

1   A. Fleischer, J. Electrochem. Soc., *94*, 289 (1948).

XVII

Mechanism of Oxygen Reduction at the Sintered Plate Cadmium Electrode.

The sintered plate nickel cadmium battery was the first rechargeable battery to be hermetically sealed. When an electrode reaches or approaches a full state of charge there is a change of electrode process from the oxidation/reduction of the active materials to an electrolysis of the electrolyte which produces a gas in aqueous systems. The gases produce a build up of pressure which can lead to rupture of the sealed container. The problem then is to solve the recombination of the gases. The early battery cell, prior to 1962, were not reliable because a small, but significant portion lost the ability to maintain a significant rate of gas recombination which resulted in explosions due to excess pressure.

The battery cells were built with the capacity limited by the quantity of electroactive nickel oxides of the positive electrode. This is easily accomplished in the manufacturing process. The quantity of electroactive material in the negative electrode was just made larger by impregnating an appropriate amount of Cd and/or $Cd(OH)_2$ into the sintered, porous structure. Using appropriate techniques the ratio of capacity in the negative electrode to that of the positive electrode was set and, furthermore, conditions set so that when the positive electrode approached full charge, the negative electrode was not near the state of full charge. The behavior is hot as simple as it might appear for a number of reasons such as the positive electrode produces some oxygen during charge, the amount of which depends on charge rate and state of charge. Also, toward the end of charged the nickel oxide the oxygen condition is an interaction with the negative electrode so that hydrogen production ensues prematurely from the $Cd/Cd(OH)_2$ electrode. This condition is avoided by making the battery cells unflooded, also known as semidry. An attempt to quantify the amount of electrolyte a method of filling using vacuum was developed. An excess amount of electrolyte was placed into the cell that had the air evacuated, the relative states of charge adjusted, the excess electrolyte removed prior to sealing. The process is described in more detail in reference[1]. By calculation of the void volume in the electrodes and the separator it was estimated that the pores were about 95% filled with electrolyte.

Since manufacture of plaque and impregnation loading levels all contribute minor variations in thickness, porosity and capacity, the use of vacuum assisted methods results in somewhat different amount of electrolyte from cell to cell, but pressure variation from cell to cell were less than the older method of placing a given amount of electrolyte per name plate capacity. In addition to a decrease of steady state pressure variations encountered in the overcharge region two other changes were encountered. One of the changes was an improvement in cycle Life. This was found to be related to air entrapped in the negative electrodes. The vacuum filling method removed the air and replaced its volume with electrolyte. The vacuum filled cells have greater amounts of electrolyte yet lower pressures during overcharge.

The other change that occurred because of the vacuum filling deals with a particular kind of auxiliary electrode called and "adsorbed hydrogen electrode". This term was made into a near acronym to make its use easier, the term "adhydrode™" was copyrighted. The adhydrodeä yielded a signal that depended upon the oxygen pressure within a sealed cell, and also upon the rate of water transport from the electrode stack to the adhydrode. Matching of adhydrode material prior to incorporation within the cells did not result in cells with matched signals. The basic cause of the differences was the fact that the original filling method introduced the same amount of electrolyte into cells of varying void volume. Using the vacuum assisted filling results in the same degree of void volume filling so that the signal was not limited by the water transport rate. The cell to cell adhydrode signal was made uniform even when stock which was not specially matched was used.Reliable maintenance free operation of hermetically sealed nickel cadmium cells was needed for the space program where long-lived orbiting satellites were used. With regard to terrestrial applications product liability was paramount.

For these reasons the mechanism of the overcharge reaction within a hermetically sealed nickel cadmium battery cell was studied.

a. Reaction Order.

The first point to be dealt with is the order of the reactions with respect to each reactant because this usually yields an insight into the process. The overall reaction being investigated is:

$$2Cd + 2H_2O + O_2 \rightarrow 2Cd(OH)_2$$

The generalized kinetic equation for this reaction is given as:

$$\frac{dp}{dt} = k' N^q p^r a_{H_2O}^s$$

where p is the partial pressure of oxygen and N is the mole fraction or state-of-charge of Cd, and a_{H2O} is the water activity. When the concentration of potassium hydroxide is not varied the a_{H2O} is incorporated in the effective velocity constant as shown here:

$$\frac{dp}{dt} = kN^q p^r$$

. The exponents q and r are the reaction orders for the two reactants, and these are the two numbers to be evaluated.

A particularly useful method for evaluating q and r in both simple and complex reactions is the variation in ratio of reactants. This is illustrated as follows:

$$\frac{\left(\frac{\Delta p}{\Delta t}\right)_2}{\left(\frac{\Delta p}{\Delta t}\right)_1} = \left(\frac{N_2}{N_1}\right)^q \left(\frac{p_2}{p_1}\right)^r$$

When the change in any reactant is not more than 10% $\frac{\Delta p}{\Delta t}$, may be substituted for $\frac{dp}{dt}$. In using this method only one of the two ratios was varied at a time. Thus the pressure decay rate may be determined at a given pressure for two or more different states-of-charge of the negative electrode. The last term becomes unity.

In the first kind of experiment the cells were fully charged, which corresponds to an 80% charge on the Cd electrodes, and the oxygen consumption rate determined. The cell was discharged 50% leaving the negatives about 40% charged[1,] and the oxygen consumption rate determined again.

The cells were pressured each time with 4 atmospheres of oxygen. If all the oxygen were consumed, the Cd electrodes would be discharged about 9% more, so that the substitution of $\frac{\Delta p}{\Delta t}$ for was justified. It was found that the ratio of rates is unity within experimental error, so that the exponent q equals zero.

The second kind of experiment was run to confirm the finding of q=0. It involved varying the state-of-charge of the negative electrode by *oxygen discharge* rather than by electrical discharge. Four cells were charged at 0.5 A, pressured to 4 atmospheres with oxygen, and the decay rates measured. The cells were repeatedly evacuated, repressurized and decays again measured. The four cells had some variation in results, but an average from this second methods of variation of state-of-charge of the negative electrode agreed with the earlier. In this latter experiment the amount of electrolyte and its distribution changed because of water consumption. This experiment substantiates the finding that q=0, the consumption rate of oxygen is independent of the state-of-charge of the negative electrode.

The decay curves showed a linearity in the semilogarithmic plot which is further evidence for the reaction being first order with respect to oxygen at pressures above 0.6 to 0.8 atmospheres.

There is still more evidence for the reaction order. A reaction vessel was made in which the amount of oxygen contained could just about fully discharge the negative electrodes. This vessel is shown in Figure XVII-1. The right side contained the counter electrodes for charging. The counter electrodes can be isolated by shutting off the valve shown.

The center compartment contained the Cd electrodes and the gas space. The left compartment was an electrolyte reservoir with plungers to adjust the electrolyte level in the cathode compartment. The test electrodes were charged to gassing, isolated, the compartment then evacuated to eliminate hydrogen. The electrolyte level was dropped by means of the plungers. There is enough electrolyte trapped on the edges to keep the electrodes wetted by capillarity. The cathode compartment was pressured with oxygen and the pressures read with the gauge or with the transducer.

The data taken in the reaction vessel at 25°C for fully charged and half-charged Cd electrodes again substantiated that the reaction is independent of the state-of-charge of the Cd electrode.

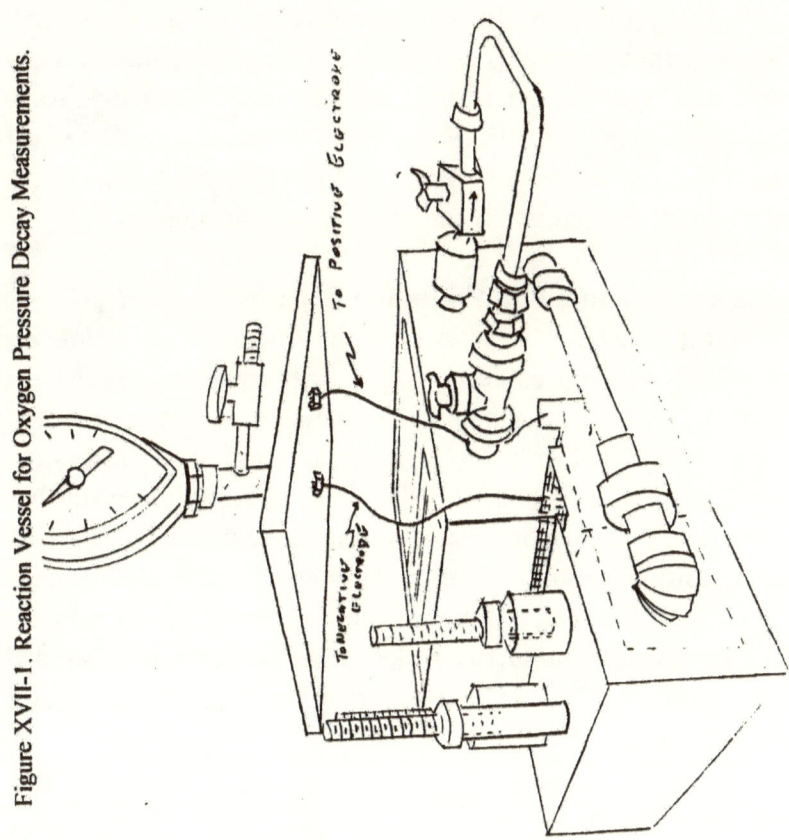

Figure XVII-1. Reaction Vessel for Oxygen Pressure Decay Measurements.

The conclusions from the several experiments based on variation in ratio of reactants are that the reaction is zero order with respect to the state-of-charge of the Cd electrode, and first order with respect to oxygen when the oxygen partial pressure is significant, above about 0.8 atmospheres.

b. Activation Energy

The activation energy for the reaction was measured in the reaction vessel of Figure XVII-1. The activation energy can serve two useful functions - (1) it can help identify the rate determining process, and (2) one can predict overcharge pressures as a function of temperature

Pressure decays were determined over the range from 25°C to 40°C. In the usual way, the logarithm of the rate constant was plotted versus the reciprocal of the absolute temperature. The straight line obtained corresponded to an activation energy of 5.08 kcal/mole.

c. Decrease of Recombining Ability and Capacity.

During the course of the work with the reaction vessel, some of the negative electrodes showed decreasing velocity constants and decreasing capacities. X-Ray diffraction patterns were made of a sintered nickel plate electrode impregnated with $Cd(OH)_2$, and of two test electrodes- one of which reduced oxygen rapidly and the other reduced oxygen slowly. The background level of the good plate (combines oxygen rapidly) is close to that of the unimpregnated sintered nickel plate, and that of the poor plate is close to that of $Cd(OH)_2$. These were interpreted by Norman Rudnick[2] as $Cd(OH)_2$ covering the surface nickel. Subsequently it was found visually that plates from button cells that develop high pressures have the white gelatinous $Cd(OH)_2$ on their surface.

d. Cadmium Foil Experiments

In 1958 Baars[3] demonstrated that the reaction sites for the consumption of oxygen at the sintered nickel plaque impregnated so as to be a cadmium/cadmium hydroxide electrode were at the elemental nickel sites still exposed. Covering of the elemental nickel by $Cd(OH)_2$ is tantamount to decreasing the number of reaction sites, and hence decrease the effective velocity constants. A Cd foil electrode was polished, washed and placed into the reaction vessel. It was prepolarized cathodically, and then its oxygen consumption rate recorded with the pressure transducer. There was an initial pressure decay that correspond to about a monolayer, based on geometric area, so that it is actually less than a monolayer when accounting for surface roughness. No further decay occurs after that. The initial decay may very well be due to adsorbed hydrogen remaining after the cathodization. When a sintered Ni plate is electrically connected to the cadmium foil and the experiment repeated, the consumption process continues. These results corroborate Baars findings that the reaction sites are on the elemental Ni sites.

e. Polarograms

Since oxygen reduction occurs at Ni sites, polarograms should also help in identifying the reaction mechanism. A polarogram with a sweep rate of 100 mV per minute was used. The current and the potential of the Ni cathode with respect to an unpolarized Hg/HgO reference electrode were recorded. Oxygen waves were not found at Ni wire nor at sintered Ni electrodes, although the concentration of oxygen does have an effect on the current.

In another sweep conducted in a N_2 atmosphere the potential was maintained constant when the electrode reached -0.92 volts, the $Cd/Cd(OH)_2$ potential. The decay of current was noted; the nitrogen flow was stopped at this point. Oxygen flow was started at this point. There was a region in which one expected a flushing of the tube and electrolyte, and the current reaching a limiting value at a higher level.

The absence of the peroxyl and hydroxyl waves are diagnostic. It should be mentioned that the perhydroxyl and hydroxyl waves were observed in 34% aqueous KOH solution at the dropping Hg electrode. Also, a separate test to detect the presence of peroxide gave negative results.

f. Theory

The mechanism of oxygen reduction at the sintered Ni electrode is important in the understanding of the sealed nickel/cadmium battery. Such an understanding leads to design of cells so that the possibility of bursting during overcharge is eliminated. The five types of experiments described above may be enumerated as follows:

1. Reaction order is first order with respect to oxygen and not dependent on the state of charge of the Cd electrode.
2. The activation energy of 5.08 kCal/mole is more likely indicative of a transport limited process than a chemical reaction.
3. The X-ray study indicates that the consumption reaction occurs on the nickel sites on the surface of the electrode.
4. The presence of Cd and $Cd(OH)_2$ tends to be a potential buffer, holding the unpoised Ni at the potential of the Cd electrode.
5. The polarographic evidence is that oxygen is not consumed via a perhydroxyl or hydroxyl process on nickel at the potential of the Cd electrode.

The mechanism proposed is that of a corrosion couple and is shown in Figure XVII-2 where the nickel site is electronically connected to the $Cd/Cd(OH)_2$ and is also ionically connected so that there is a complete electrochemical circuit. The nickel site reaches the potential of the $Cd/Cd(OH)_2$ by reacting with the water of the electrolyte while the elemental Cd sends electrons through the electronic circuit and the hydroxyl ions migrate to the Cd site to form additional $Cd(OH)_2$. These reactions are represented as:

anode: $\quad Cd + 2OH^- \rightarrow Cd(OH)_2 + 2e^-$

cathode: $\quad Ni + 2H_2O + 2e- \rightarrow 2Ni - H + 2OH^-$

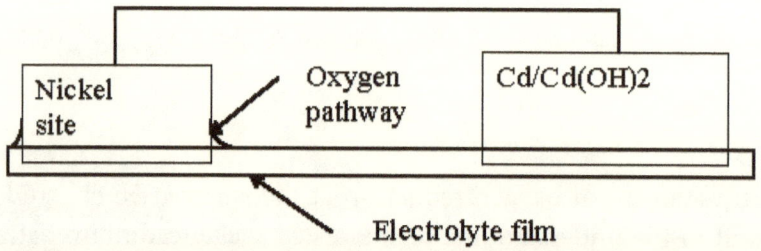

Figure XVII-2. Representation of the Proposed Reaction Mechanism for the Consumption of Oxygen at Nickel Sites on the Sintered Plate Cd Electrode.

Subsequently, when oxygen reaches the nickel site where hydrogen is adsorbed they react to form water.

chemical: $\frac{1}{2}O_2 + 2Ni - H \rightarrow H_2O$.

net: $\frac{1}{2}O_2 + Cd + H_2O \rightarrow Cd(OH)_2$.

Since the equilibrium with the potential of the $Cd/Cd(OH)_2$ is upset there will be a continuance of the above electrochemical reactions. Cd is not directly involved in the rate determining step of oxygen with adsorbed hydrogen so it meets the requirement of a zero order, while the overall reaction rate is dependent upon the pressure of oxygen. The transport of oxygen through a film of electrolyte on the nickel sites is probably the most important step which would be consistent with the low activation energy of 5.08 kcal/mole. This mechanism is also consistent with the Cd foil experiment and with the polarographs, particularly at the potential of $Cd/Cd(OH)_2$ electrode. It should be noted that actual hydrogen evolution from a sintered nickel surface does not occur until its potential reaches about -0.98 volts with respect to the SHE.

When applied to sealed battery cells, the conditions described above are applicable only on open circuit. From this point on we want to direct our efforts toward substantiation of the proposed mechanism. Consider an open (to the atmosphere) cell in which the anode is producing oxygen on overcharge and the cathode is sintered Ni which is devoid of Cd. Interrupt the current, evacuate the gases and replace them with oxygen at a pressure below the expected equilibrium pressure of the sealed system. Allow this system to rest until it is established that the pressure is constant. During this time any hydrogen on the sintered nickel would be stripped. Place the cell on charge. When the oxygen consumption rate is equal to its generation rate which is governed by the current there is a steady state. This then is our first test of the validity of the theory.

A second test for the validity of the theory is based on the chemical kinetic reaction for the rate determining step. Starting with a clean sintered nickel so that there are no hydrogen atoms adsorbed, and with some finite oxygen pressure, the pressure decay should be zero. When the sintered nickel is cathodically polarized there should be a lag in the oxygen consumption while the adsorbed hydrogen atom coverage is building up toward an equilibrium value. Even more important than the delay or lag period, when the current is interrupted, the consumption of oxygen should continue at an ever decreasing rate!

1. First Test of the Theory

Cells with nickel oxide electrodes and sintered nickel electrodes were prepared (6 Ah size). They were equipped with a valve and pressure gauge so they could be sealed in a controlled way and the pressure of the generated gases measured. A diagram of an experimental cell is shown in Figure XVII-3. They were charged vented, evacuated to eliminate hydrogen and pressured to four atmospheres with oxygen. They were then returned to overcharge at 0.6 amperes. The pressure decreased, and, after 24 hours the pressure was 1-2/3 atmospheres at 1.42 volts in one cell, and the second cell was at 1.43 volts with 2 atmospheres of pressure. Pressure equilibrium had been reached in both cases and the voltage on the c/10 charge was that usually observed in sealed nickel cadmium cells on prolonged overcharge. Cell conditions were set so this constitutes evidence that oxygen from the charged positive electrode was being consumed at Ni sites on the unimpregnated Ni sinter serving as the negative electrode at potentials the same as normal negative electrodes.

2. Second Test of the Theory

The same kind of cell was made with uncharged nickel oxide electrodes and sintered nickel plates. When charged the nickel oxide electrodes should not evolve more than an infinitesimal quantity of oxygen. The cell was pressured with oxygen and then allowed to stand on open circuit.

No pressure decay was measured. The oxygen atmosphere of the cell was decreased to ambient barometric pressure and a 1 milliliter graduated pipette attached to the valve. The pipette had a short column of water to act as an indicator for changing volume.

A 0.6 ampere charge was started. There was a lag of 10 to 20 seconds before any oxygen was consumed. The current was then interrupted and the consumption of oxygen continued for at least 90 seconds afterward. These are the results required to substantiate the theory. Hydrogen atoms at the interface of nickel/electrolyte/atmosphere react with gaseous oxygen.

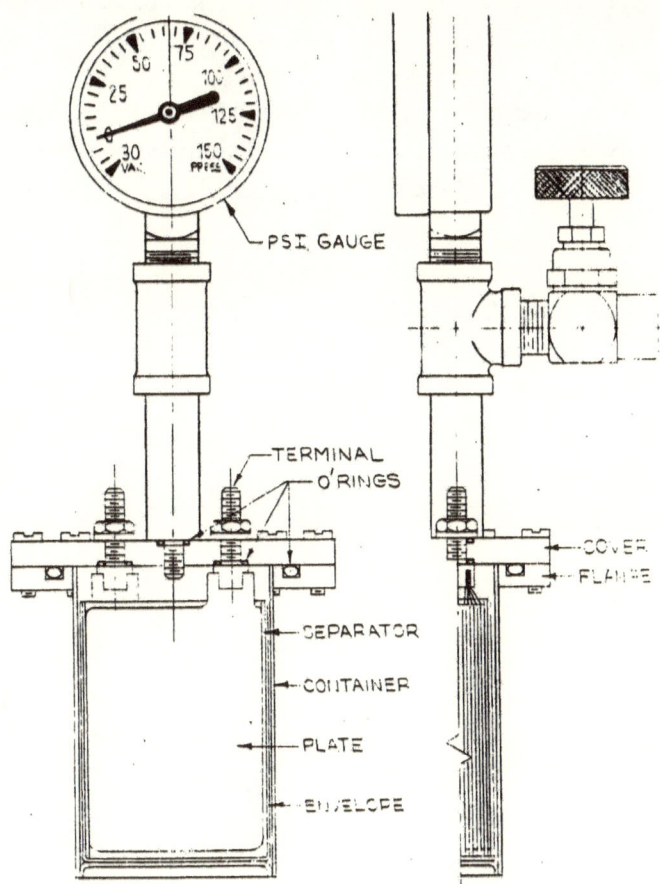

Figure XVII-3. Six Ampere-hour Sealed Nickel Cadmium Experimental Cell.

In a separate test, a three electrode system was set up. The third electrode was an unimpregnated Ni sinter. With oxygen present in the sealed cell a current was demonstrated to pass between the unimpregnated Ni sinter auxiliary electrode and the $Cd/Cd(OH)_2$ electrode. Further, the current was proportional to the oxygen pressure.

The experimental findings are consistent with the theory of oxygen recombination at the sintered plate $Cd/Cd(OH)_2$ electrode.

1 H.N. Seiger, *Battery Design and Optimization* Ed. Sidney Gross, The Electrochemical Society, Proceedings Volume 79.1.

2 Norman Rudnick, private communication.,

3 E. Baars, Paper presented at the October 1958 Meeting of the Electrochemical Society. H.N. Seiger, *Battery Design and Optimization* Ed. Sidney Gross, The Electrochemical Society, Proceedings Volume 79.1.

About the Author

Harvey N. Seiger returned to Brooklyn College at the end of WW II and received the Bachelor of Science Degree with Honors in Chemistry in 1949. Continuing on in the Graduate Extension of Brooklyn College, the Master of Arts Degree in Chemistry was awarded in 1952. Continuing on at the Polytechnic Institute of Brooklyn, the Doctor of Philosophy Degree in Chemistry was awarded in 1962. The graduate level work was all done on a part time basis while working full time and even serving as a Lecturer in Chemistry at Brooklyn College.

He became a member of Phi Lambda Upsilon (The Honorary Chemical Society). Made a Fellow of the American Institute of Chemists as well as being Accredited as a Professional Chemist by this organization. Received the "IR 100 Award" in 1967 for a Rechargeable Lithium Halide Battery, and given a NASA Certificate of Recognition in 1979.

He worked in industry becoming the Director of Research of the Battery Division of Gulton Industries until 1970. Then lead the Battery Group for Textron within the Heliotek Division until 1975. He became the Corporate Vice President of Research and Development for Yardney Electric Corporation. This was followed as a Program Manager for Electrochemistry for the Naval Systems Division of Westinghouse until retirement in 1993.

He was appointed the Secretary of the New Energy Sources Committee of the IEEE in 1960. He was a Councilor for the Los Angeles Section of the Electrochemical Society in 1970. He was also a Member of the National Battery Advisory Committee (Department of Energy) from its inception until it was disbanded.

To date he was issued 26 US. Patents . He has authored 67 papers published or presented at scientific or technical meetings which include:

- Panelist Pacific Energy Conversion Conference, 1962
- Panelist Sealed Cells Roundtable, Electrochemical Society, 1964
- Guest Lecturer Mechanical Engineering Colloquia, Polytechnic Inst. of Brooklyn, 1967
- Guest Lecturer Energy Conversion at the University of Southern California, 1974
- Guest Speaker New York/New Jersey Section of the American Institute of Chemical Engineers, 1968
- Guest Speaker Indiana Section of the Society of Automotive Engineering, 1969

He is currently listed in Who's Who in the Midwest and formerly in American Men and Women of Science, Who's Who in Science and Technology and Men of Achievement in 1981 et. Seq. He has been listed in"2000 Outstanding Scientists of the 20th Century" . He retired in 1993 and has devoted himself to answering a list of questions made during the last decade of active employment.

Appendix

The equations derived in the body for leakage currents based upon the physical model of electrochemical systems readily lend themselves to solution by matrix algebra.

Consider an electrochemical system, such as a battery, having N cells in series and having pathways consisting of channels from each cell which lead to manifolding. It was shown that the terminal electrodes, the anode at the negative end and the cathode at the positive end, do not have leakage currents. Eliminating this pair then leaves N-1 set of electrodes experiencing leakage currents. The matrix for the coefficients of the equation is square of order N-1.

The force driving leakage current is the voltage differences of the electrode pair in contiguous cells which are connected through the intercell connector. These are the anode of cell x and the cathode of cell x+1, for which one writes:

$$V_{\alpha x, \kappa(x+1)} = V_{\kappa(x+1)} - V_{\alpha x}$$ where $V_{\alpha x}$ and $V_{\kappa(x-1)}$ are oxidation-reduction potentials with respect to some reference electrode. One way to measure the driving force directly is using a reference electrode as a probe. The driving force may also be inferred from polarization data obtained from the battery itself, battery geometry and materials. The driving force results in an ionic flux, the magnitude of which is governed by the ratio of the force to the electrochemical impedance to ionic flow in the pathways to and from cells and the manifold. The channel impedance may be represented by R_c and each manifold segment by R_m. The manifold segment lies between cells, the lengths being limited by the channel entry. Thus, there are N-1 manifold segments.

Using the physical model and the equation derived in the body, the leakage currents flowing between cell are denoted as $j_{x,x+1}$. Using this symbol along with the previous definitions for channel and manifold impedance, 3 kinds of equations are found:

First Loop Equation

$$\left(2 \times R_c + R_m\right) \times j_{1,2} - R_c \times j_{2,3} = V_{k2} - V_{\alpha 1}$$

Recursion Equation

$$- R_c \times j_{x-1,x} \left(2 R_c + R_m\right) \times j_{x,x+1} - R_c \times j_{x+1,x+2} = V_{k(x+1)} - V_{\alpha x}$$

Last Loop Equation

$$- R_c \times j_{n-2,n-1} + \left(2 R_c + R_m\right) \times j_{n-1,n} = V_{kn} - V_{\alpha(n-1)}$$

which yield the following Matrix form:

$$
\begin{vmatrix}
2R_c+R_m & -R_c & 0 & . & . & .. & 0 \\
-R_c & 2R_c+R_m & -R_c & . & . & . & . \\
0 & -R_c & 2R_c+R_m & -R_c & . & . & . \\
. & .. & -R_c & . & . & . & . \\
. & . & . & . & . & . & 0 \\
. & . & . & . & . & 2R_c+R_m & -R_c \\
0 & . & .. & . & 0 & -R_c & 2R_c+R_m
\end{vmatrix}
\times
\begin{vmatrix}
j_{1,2} \\
j_{2,3} \\
. \\
. \\
. \\
. \\
j_{(n,n-1)}
\end{vmatrix}
=
\begin{vmatrix}
V_{\kappa 2}-V_{\alpha 1} \\
V_{\kappa 3}-V_{\alpha 2} \\
. \\
. \\
. \\
. \\
V_{\kappa n}-V_{\alpha(n-)}
\end{vmatrix}
$$

The array of resistance terms are the coefficients matrix which may be denoted as R. The columns of contiguous electrode voltage difference terms is the constants vector, V, leaving the column of leakage currents to be identified as the solutions vector, j. The matrix relationship is:

R×j=V

The R matrix may be factored into an upper triangular matrix, U, and a lower triangular matrix, L, where L×U×j =V , and:

$$\mathbf{L} = \begin{vmatrix} \omega_1 & \cdot & & \cdot & \cdot & \cdot & & 0 \\ \beta_2 & \omega_2 & & \cdot & \cdot & \cdot & & \cdot \\ 0 & \beta_3 & \omega_3 & \cdot & \cdot\cdot & \cdot & & \cdot \\ \cdot & \cdot & \beta_4 & \cdot\cdot & & \cdot & & \cdot \\ \cdot & \cdot & & \cdot & \cdot\cdot & \cdot & & \cdot \\ \cdot & \cdot & & \cdot & \cdot & \cdot\cdot & & \cdot \\ 0 & \cdot & & & \cdot & \cdot & \beta_{(n-1)} & \omega_{(n-1)} \end{vmatrix}$$

$$\mathbf{U} = \begin{vmatrix} 1 & \alpha_1 & \cdot & \cdot & & \cdot & & 0 \\ \cdot & 1 & \alpha_2 & \cdot & & \cdot & & \cdot \\ \cdot & \cdot & 1 & \alpha_3 & & \cdot & & \cdot \\ \cdot & \cdot & \cdot & \cdot & & \cdot & & \cdot \\ \cdot & \cdot & \cdot & \cdot & & \cdot & & \cdot \\ \cdot & \cdot & \cdot & \cdot & & \cdot\cdot & & \alpha_{(n-2)} \\ 0 & \cdot & \cdot & \cdot & & \cdot & & 1 \end{vmatrix}$$

and letting $\mathbf{U} \times \mathbf{j} = \mathbf{z}$ so that $\mathbf{L} \times \mathbf{z} = \mathbf{V}$ from which:

$$\mathbf{L} \times \mathbf{U} = \begin{vmatrix} 1 & \alpha_1 & \cdot & \cdot & \cdot & \cdot & \cdot \\ \cdot & 1 & \alpha_2 & \cdot & \cdot & \cdot & \cdot \\ \cdot & \cdot & 1 & \alpha_3 & \cdot & \cdot & \cdot \\ \cdot & \cdot & \cdot & \cdot\cdot & \cdot & \cdot & \cdot \\ \cdot & \cdot & \cdot & \cdot & \cdot & \cdot & \cdot \\ \cdot & \cdot & \cdot & \cdot & \cdot & \cdot & \alpha_{(n-1)} \\ \cdot & \cdot & \cdot & \cdot & \cdot & \cdot & 1 \end{vmatrix}$$

where the following substitutions were made:

	$\omega_1=(2R_C+R_M)+\alpha_1R_C$	$\alpha_1=(-R_C)/\omega_1 + \alpha_1R_C$
$\beta_2= -R_C$	$\omega_2=(2R_C+R_M)+ \alpha_2R_C$	$\alpha_2=(-R_C)/\omega_1 + \alpha_2R_C$
$\beta_3= -R_C$	$\omega_3=(2R_C+R_M)+ \alpha_3R_C$	$\alpha_3==(-R_C)/\omega_1+ \alpha_3R_C$
.	.	.
.	.	.
$\beta_{N-1}= -R_C$	$\omega_{N-1}=(2R_C+R_{M+})+\alpha_{N-2}R_C$	$\alpha_{N-1}==(-R_C)/\omega_1\alpha_1R_C + \alpha_{N-1}R_C$
$\beta_{n=} -R_C$	$\omega_N=(2R_C+R_M)+ \alpha_{N-1}R_C$	.

From inspection of $\mathbf{R}$ and $\mathbf{L\times U}$ it if found that the z vector is :

$$Z_3=(V_{\alpha3\kappa4}+R_CZ_2)/(\omega_1 + \alpha_2R_C)$$

.

.

.

$$Z_N=(V_{\alpha(N-1)\kappa N} + R_CZ_{N-1})/(\omega_1 + a_{N-1}R_C)$$

Once the Z vector is evaluated, the j vector can be obtained by back substitution:

$$j_N = Z_N$$
$$j_{N-1} = Z_{N-1} - \alpha_{N-1}j_N$$

.

.

$$j_{N-y} = Z_{N-y} - \alpha_{N-y}j_{N-y+1}$$

The **j** vector so far obtained is but a first approximation to the leakage current which is the object of this work. It shall now be assumed that the polarization relationship is used to estimate $V_{axk(x+1)}$ based on the total current, $I+j_{x,x+1}$ through the conjugate set of electrodes. A better or improved approximation to the constants vector is defined by the first approximation and an improved approximation to the solutions vector is obtained. Successive iterations can be done until any degree of refinement considered adequate is reached. A reasonable degree of refinement would be that $j_{n/2,n/2+1}$ changes by 0.1% or less. The first guess for $V_{axk(x+1)}$ would be V_{BAT}/N, the average cell voltage.

The equations have been written in a form suitable for program coding using BASIC. The amount of memory is small enough that small pocket computers with limited memory are suitable for the calculations.

Before encoding, we should also consider the power dissipation. The equation for the power dissipated internally (between the electrodes) and inside the battery case (channels and manifolds) was derived containing the voltages, load and leakage currents and channel and manifold imped- ances. Ail these terms were either obtained for the leakage current and voltage calculations or are a result of these calculations. Thus, the power dissipation is readily calculable. The pertinent equation is:

$$P_o = \left[\left(V^o - V_{\kappa 1 \alpha N}\right) + \sum_{x=1}^{N-1}\left(V^o - V_{\alpha x \kappa(x+1)}\right)\times\left(I + j_{x,x+1}\right)\right] + R_C\left[2\left(j_{1,2}\right) + \sum_{2}^{N-1}\left(j_{x-1,x} - j_{x,x+1}\right)\right] + R_M\left(\sum_{x=1}^{N-1} j_{x,x+1}\right)^2$$

The power delivered to the load is IV_{BAT} and by multiplying by time, the energies can be calculated, for instance:

$IV_{BAT}t$ = the energy delivered to the load,

P_Dt = the energy dissipated within the battery structure,

$(IV_{BAT}+P_D)t$ = energy discharged from the battery.

The terms $(I+j_{x,x+1})t$ are the capacities discharged from the electrodes and are useful for determining mismatch of capacity due to leakage currents.

BASIC Code for Leakage Currents

The program is dimensioned appropriately depending upon the number of cells in series, N and the computer RAM. Dimensioning reserves space in memory and can interfere with calculations. Channel and manifold impedances are denoted as C and M for encoding and are measured or calculate separately prior to running the program. The open circuit voltage and current relationships are also determined by experimentation. Lines 300 and 310 assume a linear polarization relationship with VO for the open circuit potential and S as the slope of the polarization curve. Lead acid batteries are well fitted by V = VO - S'I, but AgO/Zn and NiOOH/Cd batteries are curved at iow current densities but a linear relationship was obtained by extrapolation to a hypothetical zero current density. Alternatively, lines 300 and 310 could be substituted by a look up table or another mathematical relationship realistically representing polarization. As used in this program, the polarization data includes internal resistance as well as electrode pair current-voltage effects.

The a,b and w terms of the triangular matrices are changed to A,B and W in the coding. It is interesting to count the number of iterations required to reach a relative constancy of $j_{N/2,N/2-1}$ and this is denoted as L in line 540. If the result of two successive iterations differs by 0.1% or less, the iteration process stops. In practice L usually lies between 2 and 6. When the internal and external loads begin to match in value, L can increase in one case L reached a value of 970.

Tridiagonal Matrix

```
100  DIM B(99),W(99),A(99),Z(99),V(99),J(99)
270  INPUT "NUMBER OF CELL, N =";N
280  INPUT "CHANNEL IMPEDANCE, C =";C
290  INPUT "MANIFOLD IMPEDANCE, M =";M
300  INPUT "CELL OPEN CIRCUIT VOLTAGE, VO = ";VO
310  INPUT "CELL POLARIZATION SLOPE, S =";S
320  INPUT "LOAD CURRENT, I, =";I
330  W(1)=(2*C+M)
340  A(1)=(-C)/W(1)
350        FOR X=2 TO N-1
360            B(X)=(-C)
370            W(X)=W(1)+A(X-1)*C
380            A(X)=(-C)/(W(1)+A(X-1)*C)
```

```
390         NEXT X
400  B(N-1)=(-C)
410  W(N-1)=W(N)+A(N-2)*C
420         FOR X=1 TO N-1
430                V(X+1)=1
440         NEXT X
450  L=L+1
460  Z(1)=V(2)/W(1)
470         FOR X=2 TO N-1
480                Z(X)=(V(X+1)+Z(X-1)*C/(W(1)=A(X-1)*C
490         NEXT X
500  H=J(N/2)
510  J(N-1)=Z(N-1)
520         FOR Y=2 TO N-1
530                J(N-Y)=Z(N-Y)-A(N-Y)*J(N-Y+1)
540         NEXT Y
550  F=J(N/2)
560  VBAT=0
570         FOR X=1 TO N-1
580                V(X+1)=VO-S*(I+J(X))
590                VBAT=VBAT+V(X+1)
595  IF VBAT=N*VO-(VO-S*I) THEN 910
600         NEXT X
610  VBAT=VBAT+(VO-S*I)
620  IF ABS(H-F)/F=.001 THEN 450
630                FOR X=1 TO N/2
640                PRINT "J(";X;"";";X+1;"X+1;")=";J(X)
650         NEXT X
660  PRINT "V(2,1)=";V(2)
670  PRINT "V(";INT(N/2);"";";INT(N/2+1);")=";V(N/2)
680  PRINT "BATTERY VOLTAGE=";VBAT
682         FOR X=1 TO N-1
```

```
683  PRINT "V(";X-1);",";X;")=";V(X+1)
684        NEXT X
686  PRINT "AVERAGE CELL VOLTAGE, VAVG=";VBAT/N
690  PD=(I^2)*S
700  K=0
710        FOR X=1 TO N-1
720              K=K+(VO-V(X+1)*(I+J(X))
730        NEXT X
740  PD=PD+K
750  K=0
760        FOR X=2 TO N-1
770              K=K+(J(X-1)-J(X))^2
780        NEXT X
790  K=K+2*J(1)^2
800  PD=PD+K*C
810  K=0
820        FOR X=1 TO N-1
830              K=K+J(X)^2
840        NEXT X
850  PD=PD+K+M
860  PRINT "POWER DISSIPATED LEAKAGE=";PD;"WATTS"
870  PRINT "POWER TO LOAD=";VBAT*I;"WATTS"
880  PRINT "TOTAL POWER=";PD+VBAT*I;"WATTS"
890  PRINT "NUMBER OF ITERATIONS=";L
900  STOP
910  PRINT "DO ANOTHER DESIGN"
920  END
```

www.ingramcontent.com/pod-product-compliance
Lightning Source LLC
Chambersburg PA
CBHW031051180526
45163CB00002BA/789

* 9 7 8 0 5 9 5 1 9 4 9 5 7 *